grains

production

processing

marketing

Preface and Acknowledgments

No one book can cover grain production, marketing, and processing in any detail. The intent of this book is to trace the movement of grains from their planting to their major end uses. The book explains the function of the futures markets, where supply and demand meet in public auction. It also examines the use of futures markets as tools to insure against risks as grain moves from producer to processor to consumer. The book, then, is a basic manual to be used in understanding the complex U.S. agribusiness system.

The book was originally developed over a period of some 30 years by the men who worked with the Grain Exchange Institute at the Chicago Board of Trade, which annually offered courses in commodity trading. The editorial staff of the Chicago Board of Trade has built this text upon that foundation. Men currently working in the grain trade have continued the tradition of helping the Chicago Board of Trade fulfull its basic educational commitments through the contribution of their expertise to the writing of this text. We express our appreciation to the following people who assisted the Education staff in preparation of the 1973 edition and this revised edition.

Keith Barrett, Hennessy & Associates
Fred C. Clark, Chicago Board of Trade Weighing Department
John W. Dickson, Jr., independent trader
William D. Hagerty, Jr., Hagerty Grain Company
Jordan A. Hollander, Hollander & Feuerhaken
James A. Howard, Cargill, Incorporated
James V. Inglimo, U.S. Department of Agriculture
Elmer L. Jacobs, Continental Grain Company
George C. Jones, retired trader
Robert Judd, National Soybean Processors Association
Harvey L. Kiser, National Grain and Feed Association
Neal Kottke, Agra Trading, Inc.
Don Leavenworth, Cargill, Incorporated
Sherman L. Levin, Livestock Business Advisory Service
Robert Liebenow, Corn Refiners Association
Lawrence R. Miller, Cargill, Incorporated
Norbert Parus, Chicago Board of Trade Weighing Department
Frank Polom, Chicago Board of Trade Transportation Department
Ted Rice, Continental Grain Company
David Seagren, Agra Trading, Inc.
Howard A. Stotler, Stotler & Co.
Burt A. Townsend, Central Soya Company, Inc.
Frederich G. Uhlmann, Drexel Burnham Lambert, Incorporated
William J. Walton, General Mills, Inc.
S.A. Watson, CPC International
Paul S. Weller, National Soybean Improvement Council
Raleigh B. Wilson, Continental Grain Company
Francis D. Wolfe, Jr., Anspacher & Associates

Education and Publications Services

Editorial
editors Lloyd Besant, Dana Kellerman, Gregory Monroe

Design and Production
design, production supervisor Richard Prey
typesetting Ivana Falajlo
art production Craig Simon

40006 9-7

Contents

1

agriculture - an introduction

Brief History

Our knowledge of the beginnings of man's agricultural activities is very recent and imprecise. It comes to us not from written historical record but primarily from archaeological findings of the remains of stores of grain and fragments of primitive tools used for cultivation.

In order to feed himself and members of his family, man was forced for thousands of years to live as a nomad, following and hunting roaming wild animals. Gradually, man learned to capture and herd animals. Domestication of wild sheep and goats appears to have begun in the Middle East as early as 7000 B.C. Learning to herd and graze animals permitted small family bands to remain longer in one place and gave them free time for developing crafts. At the same time, keeping a herd of animals increased man's need for plant foods and led to his first attempts at cultivation of wild plants.

In many places, archaeological findings have included the wild varieties of some of our present cereal crops, chiefly wheat and barley. Ancient strains of wheat and barley have been found at various locations in Iraq, Iran, Palestine, and Turkey in the remains of cultures dating to 6500 B.C. It would appear that man's first attempts at cultivation of grain occurred in these areas. The grains found in later cultures appear to be types that were probably brought from these earlier Middle East locations.

On the North American continent, the cultivation of maize or corn dates from around 5000 B.C. in Mexico; there is also evidence of the cultivation of primitive forms of beans in Mexico and South America about the same time. The bean crops appear to have been an important protein element in North American man's diet, the counterpart to meat protein in the diet of Middle Eastern and European man.

At about the same time as maize cultivation was developing in Mexico, strains of barley and wheat appear to have been introduced in Africa in the Sudan and Ethiopia. Cereal crops native to Africa, principally sorghums, appear to have been first cultivated in the Sudan and Ethiopia much later, around 2500 B.C.

Cereal crops from Middle East locations are believed to have spread to Italy, central Europe, and Germany as early as 4000 B.C. Evidence of wheat and barley cultivation in France, the British Isles, and Denmark dates from approximately 3500 B.C. These advances of grain cultivation into the wooded areas of northern and western Europe coincided with the clearing of major forest lands from 3500 to 3000 B.C.

Cultivation of strains of wheat also occurred on the Indian Subcontinent as early as 4000 B.C. Archaeological evidence of the cultivation of wheat, barley, peas, sesame, cotton, and rice occurs in the period 2300 to 1750 B.C. in the cultures of West Pakistan, Kashmir, Punjab, Sind, and Rajastan. Evidence of cultivation of rice and sorghums of African strains has been found in the remains of cultures on the Indian Subcontinent ranging from 2000 to 1700 B.C.

Following the clearing of forest lands in Europe, the *cut and burn* method of cultivation was used. Farmers grew their crops, harvested them, burned the fields behind them, and moved on to new fields. While the method was inefficient, it conserved the soil. In parts of Africa and the Middle East, on the other hand, croplands were planted season after season without time for replenishment, exhausting their utility.

Agriculture's Contribution

Primitive agriculture made many contributions to man's development as a member of society, a builder, and a craftsman. It eliminated the need for the nomad hunter's wandering in search of game. It allowed for the production and storage of ample food and for population growth. With a more stable existence provided by agriculture, population grew both through increased birthrate and longevity.

In the beginning, the only social unit that could exist was the wandering nomad family. Later, however, adequate food production allowed man the time for developing crafts. With some of their food needs met by others, groups of people had some time freed to develop crafts—the beginnings of man's industrial experience. Much of their effort was applied to making crude tools to be used in cultivating plants, which made agriculture more efficient. This was the beginning of the interdependence of urban industrial and rural agricultural societies.

Conservation Practices

Although there were continuous improvements in farm tools throughout man's agricultural development, there was little innovation in the management of soil and croplands until the 18th century. In the late 1700s, in England, the Industrial Revolution greatly increased the demand for more and better food. Population was multiplying and prosperity was growing. Landlords and tenant farmers intensified the production of both field crops and livestock. Grazing sheep provided a natural source of fertilizer, and the manure from cattle wintered in yards was spread after the spring planting. Crop and livestock production complemented each other and contributed to the quality of the soil.

Today, except in the underdeveloped nations, the practice of farming has taken on the aspects of a science. Economy of scale has led increasingly to fewer but larger, more efficient, highly specialized farm units. As recently as the late 1950s, a typical farm produced a hay crop, perhaps two or three grain crops, poultry, pork, and perhaps both beef and dairy cattle. This type of farming operation is no longer economically viable. Increasingly, farming is specialized in production and scientific in the use of improved strains of livestock and plants, fertilizers, control chemicals, mechanization, and finance. Despite these remarkable changes, much is still the same. The placing of seed in the ground is just the beginning of the long grain-production/marketing process. Success is not haphazard; successful grain production begins with an understanding of soils.

Soil Management

Discussion of soil management centers on three principal factors—soil composition, soil fertility, and climate, i.e., temperature, wind, and rainfall. Soil, the outer loose covering of the dry surfaces of the earth, has been formed over millions of years by meteorological, geological, and biological actions on rock. As an example, heavier rainfall over a period of eons in one area will eventually leach out more of the soil's essential chemicals than would be leached out in a more arid climate. Excessive rainfall and runoff also leads to soil erosion. Given the variation in rainfall levels, different soil types would develop in two locations even if the parent rock material were identical in both.

Meteorological activity also influences the topography, soil structure, and soil fertility—the soil's ability to deliver essential chemicals for plant life. Most of the chemical elements that plants require for growth are in the soil in the form of cations—atoms carrying positive electrical charges. The ability of the soil to surrender these positively charged particles to the plant root hairs is called *cation exchange ability.*

Geological influences, such as glacial movements and volcanic activity, have had obvious effects on the earth's topography, soil structure, and soil fertility.

Soil Composition

One aspect of soil composition is *texture,* the size of soil particles; the other is *tilth,* the manner in which the particles are arranged.

Soil falls into one of three texture types:

sand—2 to 0.05 millimeters, diameter
silt—0.05 to 0.002 millimeters, diameter
clay—0.002 millimeters and finer, diameter

Farmers often refer to soils with higher proportion of sand as *light.* However, by actual weight per volume, so-called light, sandy soil is heavier than soil high in clay content, which is referred to as *heavy soil.* The farmer's perspective of heavy and light soils points up the fact that soil texture affects the density of its makeup and, therefore, its ability to hold water and allow air circulation. Texture also affects the ease of cultivation.

Tilth is the manner in which the particles are arranged. In good soil, particles are typically clustered into granules that create open, irregular spaces through which air and water are supplied to the plant root hairs.

Soil tilth is also important for bacteria and organisms in the soil which transfer soil nutrients to the plant's root system. The root hairs receive little of the oxygen taken in by a plant through its leaves.

Aeration, Moisture Retention

Adequate circulation of moisture and air is vital to plant growth and is affected by both the texture and tilth of the soil. The ability of soil to be worked is determined by its ability to hold water. Some soils, for example, those heavy in clay content, are sticky when wet and become hard when dried out. The ability of soil to be easily cultivated over a wide range of moisture content is to a large extent a measure of its productivity.

There is dramatic improvement in crop productivity of lowland fields that are tiled for improved drainage, due to the removal of excess water and the improvement in air circulation. On the other hand, good tilth can be ruined by puddling after flooding or by packing under the weight of heavy machinery when wet.

Fertilization

In addition to light and moisture, plant growth requires 16 essential elements. Three are readily available in air and water: carbon, hydrogen, and oxygen.

The remaining 13 must be supplied by the soil:

primary elements	nitrogen phosphorous potassium
secondary elements	calcium magnesium sulphur
micronutrient elements	boron chlorine copper iron manganese molybdenum zinc

Most soils do not have sufficient amounts of the latter 13 elements to maximize the sunlight and moisture that are available. Chemical fertilizers of various types must be added to soil to allow plants to utilize available light and moisture.

The primary elements are, by far, the ones most necessary for soils to produce maximum yields. The primary elements in fertilizers, phosphorous and potassium, are called phosphate and potash, respectively.

The secondary elements are present in sufficient amounts in many soils but deficient in others, and must be replenished by fertilization. The micronutrient elements are present in sufficient amounts in most of the world's soils and rarely need to be supplemented by fertilizers.

All of the essential elements must be kept in relative balance for soil to maintain optimum productivity. Soil scientists test soil composition frequently to detect gradual changes in the presence of essential elements that may take years to become noticeable. Scientists use both soil tests and plant tissue tests to analyze soil fertility. The soil test is a chemical analysis of the availability of essential elements. Though conducted on a less-frequent basis, plant tissue tests can uncover a complex chemical situation called "hidden hunger" in which an essential element in the soil is somehow prevented from being effectively used by the plant.

Fertilizer Designations

Fertilizers are generally designated according to the pounds of the primary element available and the form in which it is provided:

primary element	form	percentage available
nitrogen	nitrogen (N)	10%
phosphorus	phosphate or phosphoric oxide (P_2O_5)	10%
potassium	potassium oxide (K_2O)	10%

The percentages of the primary elements are always listed in the same order (1) nitrogen, (2) phosphorous, and (3) potassium. An example of a 10-10-10 fertilizer is one that contains a minimum of 10% nitrogen, 10% phosphate, and 10% potash. This same sequential designation is also used on straight fertilizer materials. For instance, normal superphosphate fertilizer, which contains 20% available phosphate, would be labeled 0-20-0. Ammonium nitrate, a fertilizer containing 33.5% available nitrogen, is designated 33.5-0-0.

Liming

With few exceptions, soil is adequate in calcium content. In these cases, lime is applied not to furnish any particular essential element to the soil, but to improve the pH, or alkaline/acid balance necessary for plants to obtain maximum nutrients from soil. Lime neutralizes the soil acidity that tends to develop as a result of the decomposition of crop stubble. Regular yearly liming is usually not necessary. In the North Central states, a two-ton application of ground limestone per acre generally will maintain the proper acid/alkaline balance for 10 to 20 years.

Control Chemicals

The use of fertilizers and, more recently, control chemicals have revolutionized agricultural production. Control chemicals—herbicides, fungicides, and insecticides—were developed to fight the specific groups of weeds, diseases, and insects that attack particular crops. The list of pesticides is so long that a general treatment of agriculture need only mention them.

The single most critical requirement for plant growth is moisture. Weeds damage plants by depriving them of moisture; controlling weeds is therefore, a basic factor in increasing crop production.

Chemical control of weeds can be used for other effects. As an example, weed killers are used on cornfields to kill all broad-leaved plants and certain weed killers are used on soybean fields to kill grasses. However, many of these herbicides leave a residue which may last for a period of years, preventing the farmer from easily rotating crops, or changing his planting plans. A soybean field that has been treated cannot easily be used for corn the next season, and a cornfield treated with a broad-leaved herbicide cannot be planted to soybeans the following season.

Many grain traders are unaware of the limitations imposed on production flexibility by these herbicides. Today, for the short term, high prices of one crop and low prices of the other will not effect changes in crop production of the two as quickly as was the case in times before the widespread use of herbicides.

Improved Seeds

Important advances in crop production have been achieved in recent years due to seed improvements made through breeding, hybridization, importing special seeds, and seed certification. The impact of hybrid seed on increased corn production, for example, has been tremendous. An example of seed improvement through importation is the introduction of various well-adapted varieties of wheat from other parts of the world. This has led to increased U.S. production of Hard Red Winter wheat in the southwestern wheat belt states. Many crops have benefited from seed certification guaranteeing that seeds meet various state requirements. Certification factors include such things as viability, the ability to germinate, freedom from objectionable weed seeds, and origin. Seeds may be treated for various reasons. Corn seed, for example, is often treated with a material that inhibits fungus organisms that attack the rather fragile corn seed before it has a chance to germinate. Legume seeds are often coated before planting with a culture containing certain bacteria that the seeds require to fix nitrogen efficiently.

Mechanization

The subject of farm mechanization should begin with some general considerations of power because the technical operations performed in agricultural production require for their adoption not so much the invention of the devices themselves as the development of an efficient source of power to drive them. The production of grain had its biggest breakthrough when draft animals were first employed as a source of power. The Romans seem not to have understood the efficient use of draft horses because their plows as well as their chariots were drawn from a belt around the horse's midsection. In the early Middle Ages, northern Europeans realized a tremendous increase in efficiency from allowing the draft horse to lean into the load by use of a collar around its shoulders. Medieval scholars believe that as a result of this invention the heavier soils of Europe were brought into production, supplying sufficient food to allow the population of Europe to double in the early Middle Ages.

The impact of the expanded use of power on modern agriculture is largely misunderstood if more stress is laid on the ease it has brought to farming than the increase in total production it has made possible. Much acreage that formerly was not profitable to cultivate has come under the plow as a result of the availability of cheaper power.

Another factor often overlooked is the contribution of speed in producing higher yields. Once the soil is warm enough to allow germination, the speed with which the entire crop can be planted determines the length of the growing season, and thus has a direct effect on yield. In 1940, the average tractor shipped by domestic manufacturers had 27 horsepower and could pull two 14-foot plows at 3 mph. In 1970 the average tractor sold in the U.S. had 100 horsepower and could pull six 14-foot plows at 6 mph.

Seedbed Preparation

Plowing, the primary tillage operation, consists of turning the top six inches of soil completely over. This operation requires the expenditure of a large amount of energy and is the single most expensive operation in grain production. Plowing is done for a number of reasons, including the removal from the surface of the residue of former crops, the loosening of the soil for other implements, the removal or delay of weed competition, and the roughing of the surface to check runoff of rainwater.

Because it lacks moving parts, the plow is thought by many to be a simple machine. However, the plow has been the subject of much detailed study and has probably given more work to agricultural engineers over the years than has any other piece of agricultural equipment. Every aspect of the plow, from the metal of the share to the type of hitch, has required much study and development to improve its efficiency. On the 18th-century plow, the *moldboard* (the curved plate that turns the soil) was made of wood, while the *plowshare* (the cutting blade), was iron. When the early pioneers pushed westward they found that the heavier lowland prairie soils stuck to the moldboard and would not slide cleanly over it. John Deere, of Moline, Illinois, is credited with developing a cast-iron moldboard that would scour itself in heavy prairie soils.

Following plowing, the seedbed often undergoes further preparation by operations known as discing and harrowing. The *disc* is an implement consisting of an arrangement of metal discs on an axle, which is pulled through the soil to slice the larger clods into smaller pieces. The *harrow* is a multitoothed device for further pulverizing the soil clumps. On many fields where a row crop has been grown and where not much crop residue is left standing, discing and harrowing may be the only tillage operations necessary.

Planting

Modern grain-planting methods are quite different from the *broadcast* or scatter, method of past generations. Small grains—wheat, barley, rye, and oats—are planted by an implement called a drill, various kinds of which have been used for many generations. The *drill* consists of a series of devices for opening up a small furrow, a tube for delivering the seed into the furrow, and a device for covering up the furrow and tamping the soil down on the seed. These devices are usually suspended below a large box used to hold the seed. Additional devices may be used to add fertilizer and other chemicals.

Before the advent of herbicides, corn was planted in hills so that the farmer could cultivate for weeds by moving across the field in two directions. Today corn is planted by a machine basically similar to a drill, although corn and soybean planters do require a more exact metering device to control the rate of planting because of the size of the seeds. The modern corn planter accomplishes a number of jobs in addition to planting the seeds, such as adding bands of fertilizers in the soil, applying herbicides and topdressing fertilizer.

Cultivation

The *cultivator* is a series of shovellike tools suspended from a bar. Each tool passes between two rows, cutting the roots of the weeds and killing them. Cultivation was formerly thought valuable for aeration of the soil as well as for weed control, but research has shown that all the benefit received is from weed control. The small grains are planted too closely for this operation.

Some years ago, corn was cultivated as many as four times, but with the development of herbicides the operations have been reduced to one or two. After the plants have reached a certain height, cultivation damages their roots, and thus herbicides have a two-fold advantage. The weeds are controlled without the expense of extra cultivation, and the roots of the plants are not damaged. If the plants develop a specific viral, bacterial, or fungus disease, or suffer from a serious insect infestation after they are too tall to allow machinery into the field, a specific treatment may be applied from the air. This type of treatment is expensive, and the threat to the crop must be serious before such treatment is considered.

Harvesting

Wheat, oats, barley, rye, and soybeans are harvested by a combined harvester thresher with a reel head that carries a sickle bar knife that cuts the plant off near the ground. A reel lays the plants on a table, over which passes an endless canvas belt that carries the plants up into the threshing device. The threshing process amounts to pounding the material on cylinders, separating the grain (kernels) from the plant by gravity, and cleaning it with a stream of air.

The corn combine has long fingerlike projections containing heavy rotating rollers that snap the ears off the plants as they pass between the fingers. The ears, with the husks, are conveyed into the thresher part of the machine. With some adjustments for each specific crop, most combines are adaptable so that the reel head may be removed and a corn head attached.

Years ago, all corn was harvested on the ear and much corn is still harvested this way today. Even the bulk of the corn to be fed to hogs and cattle was on ears. The removal of the kernels from the ear, or *shelling,* was done later in the season when shelled corn was needed for sale or feed. Corn on the ear was allowed to dry in the field before being stored to prevent spoilage because of high moisture content. However, when the high-speed rollers of a corn picker hit many of these dry and somewhat brittle ears, they snapped off easily, fell to the ground, and were lost.

The growth of combine harvesting of corn (also called *field shelling* or *picker shelling*) has been spectacular since its introduction in the 1950s. The development of combine harvesting of corn has had an important impact on grain marketing. The corn harvested by this technique is picked much earlier and at a higher moisture content than formerly, thus preventing field loss due to shattering. The combination of the increased speed of harvesting by this method and increased yields has resulted in heavy runs of high-moisture corn in a short period of time at harvest. This can be attested to by the long lines of trucks at elevators all over the corn belt during harvest season. Field shelling has made many corncribs obsolete because it requires storage structures that resist weather, insects, birds, and rodents.

Combine harvesting has made the farm dryer a new and important machine on many corn belt farms. The failure to properly control the artificial drying of corn, however, results in many broken kernels, which are classed in grading as foreign material. Increased foreign material is probably the biggest problem the grain trade has had in adjusting to combine-harvested corn. Artifical drying may also affect the yield of starch and oil in corn processing.

Use of Credit

The last aspect of crop production to be considered is the modern use of credit in agriculture. Years ago, very little credit was required. Seed was used from the previous crop; capital expenditures for lime, fertilizer, machinery, and other equipment were paid for with cash. As the size of units of production increased and the amount of capital inputs necessary for each unit expanded, the need for credit as a tool of production began to grow. Today's commercial producer may obtain financing from a number of government and private sources. The management decisions required in this area are becoming as important as those involved in the other aspects of production.

2 grain production

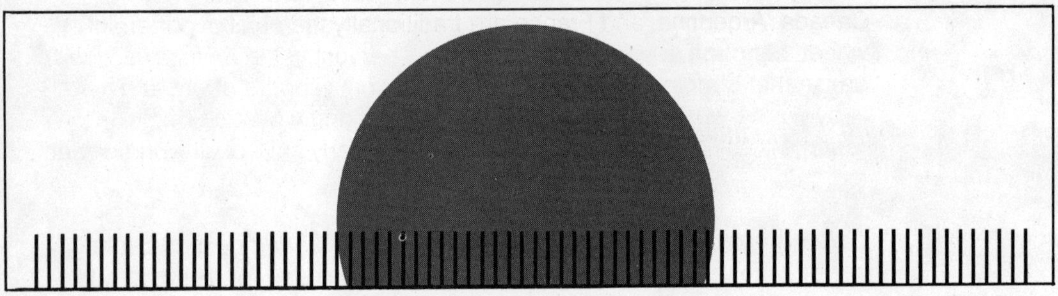

Grains are members of the grass family, the most abundant type of green plant. For purposes of convenience, grains can be divided into bread grains and feed grains. The bread grains include wheat and rye, and the feed grains (so named because a large percentage of their production is fed to animals) include corn, oats, barley, and grain sorghums. The soybean is actually a legume rather than a grain, but since this crop's growing, handling, and trading are so much like those of the grains, it is considered as a grain.

Wheat

The first wheat cultivation dates back 7,000 years. The number of wheat plants now growing on the earth probably exceeds that of any other seed-bearing plant, wild or domesticated. Wheat is grown in almost every country of North America, Europe, and Asia. Annual world production in recent years has been over 12 billion bushels. Annual U.S. production averages around 1.7 billion bushels of all classes of wheat. The U.S., Canada, Argentina, and France are traditionally the chief exporters of wheat, although in some years a plentiful harvest in the Soviet Union will cause that country to become an important net exporter of wheat. The U.S. is by far the world's largest exporter of wheat and wheat products. In recent years, U.S. exports have comprised nearly 44% of all world wheat exports.

The U.S. exports many different types of wheat; Canada exports spring wheat, whose high protein content makes it desirable for blending with wheat of lower protein content to give a higher baking quality; Australia has traditionally exported a soft wheat comparable to U.S. Soft Red Winter wheat and White wheat; while Argentina exports wheat that classifies between the U.S. Hard and Soft Winter wheats.

A. U.S. Wheat Production

Wheat is grown over much of the U.S., where the great variety of growing conditions is close to being equaled only by the Soviet Union. The crop year for U.S. wheat runs from July 1 through June 30. Wheat is harvested prior to July 1 in some areas, but the U.S. Department of Agriculture (USDA), in providing wheat statistics, reports on a July-through-June basis.

The U.S. federal grading system divides wheat into the following classes:

1. Hard Red Winter
2. Soft Red Winter
3. Hard Red Spring
4. Durum
5. Red Durum
6. White
7. Mixed

By far the most important class of wheat grown in the U.S. is Hard Red Winter, whose center of production is the western two-thirds of Kansas and adjoining parts of Nebraska, Oklahoma, and the Panhandle region of Texas. In these areas the annual precipitation averages less than 25 inches, with frequent dry periods, and the winters are often accompanied by subzero temperatures.

The Soft Red Winter wheat belt extends from central Texas northeastward to the Great Lakes and then east to the Atlantic Coast. With the exception of the western edge of the belt, Soft Red Winter wheat is grown in a subhumid climate that is not conducive to the production of wheat of hard kernel texture.

Hard Red Spring wheat is grown principally in the North Central states where the winters are too severe for winter wheat production. The principal Hard Red Spring wheat districts have a fairly deep black soil and dry, hot summers—both important factors in the production of a high-grade wheat suitable for milling into bread flour.

Durum wheat is grown in California, the Dakotas, Minnesota, and Montana. Durum and Red Durum differ from other wheat produced in the U.S. Durum flours are called *semolina* and are used for spaghetti, macaroni, and similar pasta products, and not for bread, cakes, or pastries. The Durum kernels are extremely hard and vitreous and contain the highest percentage of protein of any of the wheat classes. Durum is naturally resistant to *rusts,* which are plant diseases caused by fungi and evidenced by discoloration of the stem. When wheat is infected by rusts, its yield is reduced.

The main producing areas for White wheat are in southern Michigan, in western New York, and on the West Coast. Its usage is similar to that of the Soft Red Winter class. On the Pacific Coast, White wheat is the most important class exported.

Each class of wheat has a special use; when mixture of classes occur, usefulness may be impaired. The federal grading system classifies wheat containing 10% or more of another class as Mixed Wheat, noting the percentage of each class for the guidance of both buyer and seller. A mixture sometimes occurs on the farm when winter-killed wheat acreage is reseeded in the spring. However, some winter wheat plants survive and are then harvested with the spring wheat. Mixtures of classes can also occur along the marketing channel.

Many varieties of wheat may be sown either in spring or fall, although some winter varieties must go through a dormant period to produce kernels. Winter wheat is usually seeded in September or early October, whenever the soil has sufficient moisture to germinate the seed. At this time of the year, the danger from insects, such as the hessian fly, is considerably less than in late summer. As the wheat grows to a good grasslike stand of 5 to 7 inches, these fields can be grazed by animals.

With freezing temperatures, winter wheat goes into a period of dormancy until warm weather returns. Ideally, a blanket of snow should cover the fields to insulate and protect the plants. If adequate snow cover is not present, wheat sometimes *heaves*—the stem is severed from the root system due to alternate freezing and thawing of the soil.

In the spring, the head, which contains the kernels, develops on top of the stem. After the kernels have fully developed and filled, the color of the plant begins to change from green to gold. When the heads of grain have dried sufficiently, the fields are ready to be combined, a process beginning in late May and usually completed by the middle of July.

In the northern part of the U.S. where winters are too severe for winter wheat, spring wheat is planted as early as possible in the spring. The dormancy period is eliminated, the crop matures in August, and combining is usually completed by early September.

B. Disposition of U.S. Wheat

Approximately one-half of the U.S. wheat crop is normally milled into flour. U.S. per capita flour consumption has declined over the years but now appears to be leveling off at around 109 pounds per person. Population growth, however, has tended to offset lower per capita consumption.

Wheat exports have traditionally been a market-balancing item. In years of large production, wheat prices have tended to decline to a level that encourages the export of the domestic surplus. Likewise, when other wheat-producing and -exporting countries produce smaller than anticipated crops, demand for U.S. wheat abroad increases.

A portion of the current year's production is saved for seed for planting the following year, and a small part of the crop is fed to livestock. Usually the quantity used as feed is relatively small because wheat is priced as a bread grain for human consumption. In years of low-quality production, however, a greater portion may be fed to livestock to avoid a total loss if the wheat cannot be used for human food.

Corn

Researchers generally conclude that corn evolved from a wild grass that grew about 60,000 years ago. A grass called *teosinte,* found today in Central America and Mexico, may be corn's ancestor. Just as wheat was the backbone of early civilization in Europe, the Near East, and elsewhere, corn was the backbone of the tribal civilizations on the American continents. The Mayan, Aztecan, and Incan civilizations all depended on corn for food and as an integral part of their religion and art. The North American Indians, as well as the colonists, relied on corn for survival. After Columbus discovered corn in the Western Hemisphere, he introduced it to Europe, where for years it was grown only as a curiosity.

Corn is now grown in almost all temperate and tropical regions of the world, in conditions ranging from 58° north to 40° south latitude; from below sea level to 12,000 feet above sea level; from the arid and semidry plains of the U.S.S.R. and North Africa to the subtropical regions of Florida; from as little as 8 inches annual rainfall in Morocco to over 200 inches in parts of India; and from short growing seasons as in Quebec to nine-month seasons as in tropical Colombia.

World corn production consists of several different kinds of corn. A familiar type is sweet corn, which is available to the consumer either fresh, canned, or frozen. Another familiar corn is popcorn, which pops when heat causes the water in the kernels to expand to steam and burst the kernel. Only about 1% of U.S. corn acreage is given to sweet corn production, however, and less than 0.3% is given to popcorn production.

About 95 percent of all corn grown in the U.S. is dent corn, which is named for the indentation in the corn kernel caused by shrinkage of the starch. Dent corn is grown extensively in the U.S. Midwest and is traded in both the cash and futures markets. Federal grain standards have classified dent corn into Yellow, White, and Mixed. Yellow by far overshadows all other classifications in the commercial grain markets, but White corn is preferred by some dry millers. Mixed corn is seldom seen.

Several other types of corn are little seen in the U.S. Flour (or soft) corn contains starch and is easily ground into meal. This kind of corn was popular with American Indians but is no longer grown commercially. Pod corn has no economic value since each kernel on the ear is surrounded by short husks. Flint corn, commonly grown in Argentina and South Africa, has a hard, or flinty, starchy layer covering the soft starch in the center of the kernel. Not much production of this type of corn occurs in the U.S., although it is preferred by some European millers. Waxy corn has a different chemical structure than other kinds of corn and is desired by some wet corn millers. Most of the waxy corn acreage in the U.S. is grown under contracts between a farmer and a miller.

A. U.S. Corn Production and Disposition

While corn is grown in nearly every state of the U.S., production is concentrated in the central portion of the country called the corn belt. Planting in the corn belt gets under way by late April and is completed in northern areas by late May. Corn planting in the southern states begins in March. In three weeks, the growing plant pierces the soil surface and, from that point, plant growth is rapid. Large amounts of moisture and soil nutrients are required. As the plant grows, the ears start to form, with each plant developing one or two ears. Maximum growth is achieved in the main producing areas by late August. After this point, the plant begins to lose its excess moisture, and its color changes from rich green to straw. When the kernels on the ear have lost sufficient moisture, picking begins, usually in early October in the main producing areas. By November 15, most of the corn has been harvested.

During the past five years, approximately 65% of the U.S. corn crop has been used as a livestock feed, and exports have represented approximately 17% in these years. Other uses for corn include industrial products, products for human consumption, and seed. *Wet corn milling* (separation of the parts of the kernel after steeping in water) is the largest processing use, producing starches, sugars, syrups, oils, and animal feeds. Statistics on *dry corn milling* (grinding or breaking the corn kernel) are not separately reported but are estimated at about half of the wet corn milling use.

B. Hybridization

A *corn hybrid* is the first-generation cross between two or more inbreds. An *inbred strain* is produced when the ear shoots of corn plants are covered with inverted paper bags so that no stray pollen from other plants can fall on the silks. At the proper time, these bags are removed and the silks are pollinated by hand from the tassels of the same plant. Bags are then replaced on the ear shoots to prevent further pollination from unwanted outsiders. Such inbred corn has runty, ill-shaped ears.

The best ears are saved for the next year's seed, and the self-pollination process is repeated—the pollen from the tassels is again dusted on the silks of the same plant. After several generations, the inbred ears have become uniform but small, and the good traits of the inbreds are firmly fixed. Crossing two such inbreds to produce a hybrid results, surprisingly, in the growth of a large and vigorous plant.

In *open pollination,* the method used prior to hybridization, a corn plant is fertilized by pollen from the tassels of other corn plants. This pollen may be blown on the wind for some distance before it drops on the silks of the female parent plant. If the pollen is from an inferior plant, some of the resulting kernels will be poor in seed quality.

Oats

Oats are thought to have been first cultivated by ancient Slavonic peoples during the Bronze Age and the Iron Age, before the Christian Era; historical evidence shows that oats were being grown for grain in Europe and used for animal feed in Asia Minor early in the Christian Era.

Oats are now grown in practically every country of the world. The best results are obtained, however, in the more temperate regions because the oat plants need plenty of moisture and cool weather, especially cool nights at the time of maturity.

Total world oats production has run from 3.3 to 4 billion bushels in recent years, with U.S. production accounting for between one-third and one-half of the world total. Other important production areas include Europe and the U.S.S.R. Oats are grown in most states in the U.S., with the heaviest production in the North Central area. The crop year for U.S. oats runs from July 1 through June 30.

Color is the most important factor in classifying oats. The federal grading system uses the following classes for oats:

1. White
2. Red
3. Gray
4. Black
5. Mixed

White oats comprise most of the U.S. production and are grown generally north of the Ohio River and westward to the Rocky Mountains. Red oats are raised south of the Ohio River, into Texas, Oklahoma, and Kansas. Gray oats are grown in the Pacific Northwest. Black oats are grown in small amounts in scattered areas throughout the country. Mixtures of these types are called Mixed oats.

Oats, like wheat and rye, have either a spring or winter growing habit. Red and Gray oats are planted in the fall and closely parallel the growth/dormancy/growth cycle of winter wheat. White oats are planted in the spring and approximate the growth cycle, appearance, and farming practices of spring wheat. Harvesting begins in late May in the South and is completed in August in northern states.

Approximately 85% of the U.S. oats crop is fed to livestock. Oats are an excellent feed for horses, breeding animals, young stock, and poultry. About 3% of the crop is used for rolled oats, oatmeal, ready-to-eat breakfast cereals, and oat flour. Seed consumes 7% of the crop, and exports represent a small percentage of total production.

Soybeans

Soybeans were cultivated as long ago as 2800 B.C. in China, where the soybean was considered one of the five sacred grains. The soybean was known in Europe by the seventeenth century and was introduced in the U.S. around 1800, although it was largely ignored in the U.S. until late in the century when several USDA experimental stations began intensive soybean studies. In the late 1920's, W.J. Morse of the USDA traveled to China and sent many soybean specimens back to the U.S. Research by the USDA and by the seed industry developed soybean varieties that flourish in the U.S.

The U.S. is now the largest producer of soybeans, growing about 1.3 billion bushels per year, some 80% of average annual world production. The crop year for U.S. soybeans runs from September 1 through August 31. Of the more than 150 varieties of soybeans grown in the U.S., adapted to a range of soil and climate conditions from Mississippi to the Canadian border, the dominant class in commercial markets is the Yellow soybean. The soybean belt covers the central and southern areas of the corn belt. With the exception of production in Mississippi, Arkansas, and along the East Coast, most soybeans are grown in the Midwest.

In the main part of the soybean belt, soybeans are planted after corn, in May or early June. Germination and soil penetration usually take place in about three weeks. The soybean plant is classified as a *legume* because it has the property of extracting nitrogen from the air for use in its own growth and for storage in the soil. As growth progresses, a bush is formed, which may grow waist-high. A plant may have 60 to 80 pods, each of which usually contains three soybeans.

Soybeans are strongly affected by *photoperiodism,* i.e., maturity is determined by the number of hours of light per day rather than by a specific number of days after germination. Therefore, a given variety of soybeans matures within a few days of a particular calendar date each year at each latitude regardless of the time of planting, although yields from late-planted fields are usually lower. In the main production areas, the lush green color begins to yellow and leaves begin to drop off in late August. After all the leaves have dropped off, combines move through the fields to harvest the crop.

About two-thirds of the total soybean crop is processed domestically into soybean oil and meal. Most soybean oil is used in human food products such as shortening, margarine, cooking oils, mayonnaise, and salad oils; about 8% is used industrially in paints, printing inks, core oils, and other products. Soybean meal is used as a prime high-protein ingredient in many livestock feeds.

U.S. exports of soybeans have increased dramatically in the last 15 years, making soybeans the nation's leading agricultural dollar-earner abroad. In 1974, foreign demand for U.S. soybeans and soybean products accounted for more than $5 billion in export sales. Much of this demand is the result of increasing desire for meat in Europe and Japan. Most soybean exports are processed abroad into oil and meal. In addition, some 20% of U.S.-produced soybean meal is exported to be used as animal feed for meat production.

3

cash and futures trading

Commodity markets consist of two large markets—cash and futures—that interact and are mutually dependent. Grain traders are concerned with both markets and with the interaction between them.

Elements of Cash Contracts

Most individuals transact cash contracts when they purchase items in stores. For example, when someone buys a loaf of bread in a supermarket, he is party to a contract with the store owner who has provided a loaf of bread in exchange for money. If the store had no bread or the buyer had no money, no contract could have been entered into. This example and all cash contracts include the following elements:

1. A contract may be consummated between any *responsible buyer and seller* at any time it is convenient to both parties.

2. A *specified quantity,* mutually agreed upon, is delivered and received.

3. The *quality is specified.* Lower or higher quality costs less or more, as agreed. For instance, a loaf of bread is assumed to be fresh today, but day-old bread would be sold at a lower price, and a specialty loaf would sell for more than the ordinary loaf of fresh bread.

4. The *time of delivery is specified.* In the example above, the time element is immediate or spot. Cash contracts, however, can also be made for any delivery time agreeable to both parties, such as today, next week, next month, and so forth. For instance, one could buy a chair and have it delivered later.

5. *The price is specified.* The price of bread, for example, is probably stamped on the wrapper.

6. *The cash contract is normally settled by full performance on the part of both parties.* It would be rare indeed for the store owner in the above situation to decide not to sell the loaf of bread to a person with the money to buy it; likewise, people normally pay for what they purchase.

In grain trading, the cash contract is more sophisticated than is the simple purchase of a loaf of bread, although the same elements of a cash contract exists in both cases. To understand the cash contract in terms of grain trading, consider the following example. Assume a cash contract exists between a terminal elevator in Chicago and an export firm in Baltimore, Maryland. The contract provides that the terminal elevator is the seller of 100,000 bushels of No. 2 Hard Red Winter wheat at $3.55 per bushel to be shipped by rail from Chicago to Baltimore during the month of July. The elements of the contract are fulfilled in this example as follows:

1. A contract entered into between any *responsible buyer and seller* at any time convenient to both parties—the buyer and seller have probably traded millions of bushels over a period of years. It is the responsibility of each party, however, to be aware of the financial responsibility of the other. If the firm is relatively new, a routine inquiry through banking connections will elicit the required information. In some cases, reciprocal margins are put up.

2. *Delivery and receipt of a specified quantity*—in this case, 100,000 bushels are specified.

3. *A specified quality*—lower or higher quality will cost less or more. The specific quality in this cash contract is No. 2 Hard Red Winter wheat. If No. 3 Hard Red Winter wheat had been specified, the price might have been only $3.50. Conversely, if No. 1 Hard Red Winter wheat had been contracted for, the price might have been $3.59. Frequently, contracts will be written providing predetermined premiums for grain that grades better than the contract grade and predetermined discounts for grain that grades lower than the contract grade.

4. *A specified time of delivery*—the entire 100,000 bushels must be shipped sometime in July. All of it might be loaded and shipped on July 1. Or it might be loaded for a series of partial shipments at different times. In either case, complete shipment must be made during the month of July.

5. *A specified price*—in this example, $3.55 per bushel is payable by the buyer to the seller upon receipt of the required documents of transfer.

6. *A settlement by full performance on the part of both parties*—In probably 99% of cash contracts, full performance (satisfaction of all contractual terms and conditions) takes place. In the infrequent event that one term or condition has not been satisfied, cash contracts themselves provide arbitration procedures, which usually prevent litigation.

Elements of Futures Contracts

A *futures contract* is a contract to deliver, if the seller elects, or to receive, if the buyer elects, a specified quantity of a particular commodity during a specified period of time in the future. The futures contract evolved from a type of cash contract for deferred shipment called a *forward contract*. The elements of futures contracts include the following:

1. *A futures contract may be entered into between a buyer and a seller who are members of an organized exchange.* The buyer and the seller are members who may be acting either for themselves or for customers. The contract is transacted at the exchange during specified hours as set by the rules and regulations. Financial responsibility is a condition of membership and is a matter of constant scrutiny by the exchange itself, as well as by the Commodity Futures Trading Commission, a federal agency regulating futures trading.

2. *Futures contracts are in specified quantities.* Organized exchanges have set the units of trade for futures contracts.

3. *Futures contracts are for a specified quality.* Exchange regulations governing deliveries against futures contracts spell out very specifically the grades, or quality, of the commodity to be delivered. Some of these grades are deliverable at the futures contract price, while other grades are deliverable at discounts or premiums to the futures contract price.

4. *Futures contracts require delivery specifications.* The time of delivery is at the discretion of the seller (the short), but delivery must be made sometime between the first business day of the current delivery month (the

month in which the futures contract matures) and the last business day of that month.

5. *Prices of futures contracts are determined by public auction.* While the price of a cash contract is subject only to negotiations between the two parties of the contract, all trades of futures contracts must be competitive and transacted by open outcry auction on the trading floor of an exchange. (An exception is the practice of *versus cash* transactions which are discussed in the following chapter.) Additionally, when a futures trade is made, the price must be reported to the exchange officials for dissemination to the public and for permanent record; the specifications and price of a cash contract remain private.

6. *Futures contracts are usually not settled by delivery.* The volume of futures contracts settled by delivery is relatively small because of the principle of offset that allows either the original buyer or the original seller to close out his original trade with an opposite offsetting sale or purchase. This is possible because the clearinghouse of the exchange assumes the other side of each trade transacted during the day. In the event a trader stays long futures, that is, has purchased futures and continues to own them after the expiration of trading in that futures month, he receives a warehouse receipt or shipping certificate in satisfaction of his position. He then makes full payment for the commodity being delivered. Conversely, a trader with a short position, who fails to offset his position with an equal purchase of futures prior to the expiration of the contract, must fulfill the contract by delivering warehouse receipts or shipping certificates for the commodity to satisfy his short position.

Types of Speculators in the Futures Market

Users of a futures market can be divided into two general types: commercial and speculative. Commercial traders are usually engaged in hedging, which is discussed in the following chapter. The several kinds of speculative traders include scalpers, day traders, position traders, and spreaders.

A. Scalper

The scalper is an exchange member and professional trader who attempts to buy at a bid price and sell at an asking price. He makes a profit if he buys and then sells higher, or sells and then buys lower offsetting contracts. The scalper's activities help to maintain market liquidity necessary for the efficient operation of the market. He will buy and sell on minimum price fluctuations trading in and out of thousands of bushels of grain a day, content to take small profits and, conversely, small losses. He usually maintains no position from one day to the next.

B. Day Trader

The day trader usually tries to capitalize on price changes that are greater than the minimum fluctuation. He attempts to anticipate price advances and declines. If he is wrong, he will usually take the loss quickly; he is content to take profits on fractional gains and usually prefers to be even at the end of the day.

C. Position Trader

A position trader carries long or short positions from one day to another. Short-term traders carry positions about one week to about one month, while long-term traders may take positions extending over a year. The position trader tries to forecast significant price movements and is therefore not interested in the minor movements that the scalper trades.

D. Spreader

The spreader is a professional trader who hopes to profit on changes in the relationship between delivery months of the same commodity, between different commodities, or between the same commodity on different markets. He simultaneously buys and sells futures contracts for the same or equivalent items on the same or different markets in order to profit from price discrepancies. The spreader is not particularly interested in whether the market goes higher or lower. He is interested in the relative differences existing between futures contracts. He is long one and short another at the same time. This requires considerable knowledge of all the elements of the grain industry. While the profit potential is not as great as in position trading, the risks are less. An additional advantage of spreading is that margin requirements are less than for an outright long or short position. Market liquidity benefits from the spreader's activities because when a heavy concentration of buying or selling orders causes the differences between futures deliveries to get out of line, the spreader buys or sells the months that are out of their normal price relationship, and thus normal differences are quickly returned.

The following examples will help clarify the spreader's activities. Assume that generally throughout the fall harvest period the December corn futures trade at 6 cents under the March. At one point in a trading session, the December contract is trading at $2.74 and the March at $2.80. Assume that a Japanese grain merchant has just booked several cargoes of U.S. corn. The fast, efficient international trading network is set in motion and results in a broker for an American exporter, who wants to hedge the sale, receiving an order in the corn pit to buy a million bushels of December corn at the market. Other brokers may have December corn for sale at $2.74½ but probably not enough will be offered at that exact minute to fill the buy order. The broker would then be required to bid higher and higher to buy all the December futures necessary to fill his order. This usually isn't necessary because of the number of spreaders in the pit.

Spreaders instantly see that the exporter's broker has bid $2.74½ for December, while over in the March area of the pit, other brokers are offering the March at $2.80. They feel that the overall pattern of the market has been for the December to be worth only 6 cents under the March and not 5½ cents under. They instantly turn to those trading the March and buy from those brokers offering it at $2.80. They then turn around and sell the exporter his December at $2.74½. These spreaders are now short December 5½ cents under the March. They believe the spread will widen to 6 cents giving them a ½-cent profit.

Assume that later in the same session the general level of futures prices has declined but the spread differences have returned to a normal pattern. December corn is now trading at $2.73 and the March at $2.79. The spreaders now have the opportunity to offset their spread and make a ½-cent profit. Somewhere, to continue our example, a number of country grain dealers decide that because the futures market will soon close for the day they had better sell futures equal to the quantity of cash corn they have bought from farmers during that morning. The apparatus is set in motion and orders to sell the December futures come into the corn pit. The December futures brokers begin to offer the futures at $2.73, but buying orders at that price do not occur at that precise moment in the pit. These selling orders would have to be executed at a lower price if the spreaders did not instantly see that the March contract is trading at $2.79. They agree to buy the December at $2.73 from those pit brokers representing the country grain dealers and sell the March contracts at $2.79. They have now bought December 6 cents under the March. The accounting looks like this:

December corn	March corn
sell at $2.74½ per bushel	buy at $2.80 per bushel
buy at $2.73 per bushel	sell at $2.79 per bushel
result: 1½ cent per bushel gain	1 cent per bushel loss

½-cent gain on spread

A profit of ½ cent on 5,000 bushels, the size of a grain futures contract, is $25.00. Most of this type of spreading is done in large volume, and, on a busy day, one of these spreaders may trade several million bushels. If the volume is large enough, a skilled spreader can earn excellent profits on his ability to react to changes with little risk. For want of a better name, we might refer to these traders as *scalp spreaders.* They usually trade for only small fractions on large volume and usually hold no position overnight.

Other spreaders take a position with a longer view. Suppose, as in our example, that December corn is traded at 6 cents under the March futures. As the harvest progresses, the spreader may become convinced that because of an increasing supply of corn coming into Chicago and because the Great Lakes will soon be closed to navigation for the winter, the December would later sell at a further discount to the March. This trader would sell the December at $2.74, and buy the March at $2.80. Assume that he had been correct in his forecast that the spread would widen, even with an advance in the market. Now March corn is trading at $2.88 and the December at $2.81. He unwinds his spread by buying the December at $2.81 and selling the March at $2.88. He may be said to have sold December 6 cents under March and unwound at 7 cents under.

December corn	March corn
sell at $2.74 per bushel	buy at $2.80 per bushel
buy at $2.81 per bushel	sell at $2.88 per bushel
result: 7 cents per bushel loss	8 cents per bushel gain

1-cent gain on spread

It should be clear by now that the spreader has no interest in the outright, or flat, price levels of commodities and the various delivery months. He is interested only in the differences or relationships among prices.

Carrying Charges

Carrying charges are based on the cost of carrying a commodity from one time period (usually harvest) until it is needed sometime in the future. In the normal carrying-charge situation, the price of the nearby or closest delivery month is the lowest, while the later or deferred months are progressively higher. An inverted carrying-charge situation is just the opposite. Inverted carrying charges reflect a shortage situation in which the price of the nearest future is higher than the price of later delivery months. Although carrying charges can be calculated for the various commodities, there is no arithmetical method for calculating the extent of an inverted market. Determination of the extent of the inversion depends on how severe the shortage situation becomes. The extent of the inversion is dependent upon economic factors that influence how high premium prices must be in order to ration use of the commodity for only the most urgent needs.

The carrying charges for grain and soybeans consist of three elements—storage, interest, and insurance. Storage is a cost of grain ownership because grains and soybeans must be protected from the weather, rodents, and insects. Interest is also a cost of commodity ownership, since money is nearly always borrowed against the inventory and therefore is a matter of individual negotiation between the borrower and his banker. Usually the interest rate on hedged grain inventory loans or warehouse receipt loans will be no higher than one-half point over the prime rate, the rate charged large borrowers with a favorable credit rating. Insurance rates on grain and soybeans vary with the insurance risk and rating of the individual elevator. Although subject to periodic revision, current insurance rates for corn stored in a large modern elevator in Chicago would be around $.000014 per bushel, per day; wheat, roughly $.000018; and soybeans, $.000030. When these individual elements are added together, the sum sets the maximum theoretical difference between any two futures delivery months in a carrying-charge situation.

As an example, assume December corn futures are $2.50 per bushel. Since, for the purpose of calculation, it must be assumed that delivery against the long position will occur at the earliest possible time, in this example, delivery would be on December 1. Additionally, the earliest possible delivery in the next futures delivery month has to be assumed; that would be March 1. Therefore the corn has to be carried from December 1 to March 1. There are 31 days in December, 31 days in January, and 28 days in February, for a total of 90 days. It is practical to figure the highest daily rate, as it might cost that much. To calculate the storage element, then, multiply the number of days the corn must be carried (90) times the storage rate per bushel per day (say, $.001) and this gives $.09 or 9 cents for the period.

The insurance element will vary depending on elevator experience. Since insurance is a small portion of the total carrying charge cost, an approximation of value is sufficient. Assume the prime interest rate is 7½% and that money can be borrowed at 8% (per annum).

A summary of these calculations is as follows:

storage	90 days × \$.001 =	9 cents per bushel
insurance	90 days × \$.000014 =	.126 cent per bushel
interest	$\frac{8\% \times \$2.50}{4}$ =	5 cents per bushel
full carrying charge	=	14.126 cents per bushel

Only infrequently do the differences between delivery months go to a full carrying charge. The reason is that spreaders and commercial interests would be very willing to buy the discount month (the December in this example) and sell the premium month (March) as they approach full carrying charges with, for example, prices at \$2.50 for December and \$2.61½ for March. They would do this knowing full well that little or no loss could be suffered.

More commonly, delivery months tend to reflect a carrying-charge situation, but not the full cost of carrying. A carrying-charge market, then, usually indicates that a commodity is seeking storage and that plentiful supplies for all delivery periods are anticipated.

Calculations of carrying charges are given careful consideration by both commercial and speculative interests in grain trading. Such calculations are among the tools used to analyze the relationship between cash and futures prices. (Carrying charges are discussed more thoroughly in Chapter 4.)

4 hedging in futures

Futures markets are used by various types of commercial commodities interests as a means of protecting their holdings or commitments of commodities. *Hedging* is a price protection or insurance device employed for the purpose of minimizing losses and protecting profits in commodity producing, storing, processing, or marketing. If cash purchases and sales of a commodity could be accomplished within a relatively short time period (perhaps a day or two), in most cases the transactions could probably be accomplished with profit, since the possibility of unprofitable price swings would be minimized by the short amount of time during which the cash commodity was held. As a commodity is held for longer periods of time, however, or as commitments requiring the commodity must be made prior to purchases, the cash position of the owner is open to loss due to price fluctuation unless he has counterbalanced his position in some manner.

Futures markets provide a counterbalancing vehicle for the protection of price and profit margin. Hedging with futures contracts counterbalances one's position in the actual commodity with an opposite and equivalent position in futures contracts. A hedger enters the futures market to stabilize his profit structure and reduce his business risk. While profits on hedging activities are welcome, this is a secondary consideration, since the primary motivation behind hedging is to substantially lessen the risk of loss. By not taking advantage of the hedging opportunities afforded by the futures market, a producer, distributor, processor, or user of a commodity is *speculating* in the cash market. If one has adequate capital and is willing to risk it in the cash marketplace, then hedging with futures may not appeal to him, but if one wants to minimize risk, futures markets give him that opportunity.

The reason that hedging cash positions with futures positions is effective is that cash and futures prices have a tendency to move in concert with each other and maintain predictable relationships in situations of fairly normal supply and demand. This parallel price movement manifests itself because both the cash and futures markets are influenced by the same price-making factors. By counterbalancing his cash position with an opposite and equivalent position in the futures market, a hedger replaces the risk of price fluctuation with risks of a change in the relationship between the cash price and the futures price of a commodity. Because the cash and futures prices tend to fluctuate together in response to shifting supply and demand factors, the difference between the two prices tends to remain relatively stable. The risks of change in the difference between the cash prices and the futures prices are much less than the risks of change in the price of either the cash or the futures.

The Basis: Relationship between Cash and Futures Prices

The difference between the cash price of a commodity at a particular time and place and the price of that commodity in the futures market on a given exchange is referred to as the *basis.* The basis is expressed as the arithmetic difference between the cash price and the futures price being quoted. For example, if a country elevator in central Illinois is buying soybeans for $5.15 per bushel on a certain day in October and the price of the November soybean futures on the Chicago Board of Trade is trading at $5.25 on that day, the basis at the country point would be 10 cents under the November. The basis in another location on that day would be 6 cents under if the cash price at that point were $5.19 per bushel.

The difference between the cash price and the futures price is due to several factors, including:

- availability and cost of transportation between the location of the cash commodity and the futures market
- supply and demand conditions at the location of the cash commodity relative to those prevailing at the futures market
- variations in quality factors between the cash commodity and the contract grade of the commodity in the futures market
- availability of storage at the location of the cash commodity relative to that at the futures market
- relative supply and demand and price levels of substitutable commodities

The basis can be positive or negative, depending upon whether the cash is under or over the futures price. Futures prices are generally higher than cash prices, reflecting the cost of carrying the commodity for future delivery. Assuming the cash price is below the futures, a narrow difference is called a *strong basis,* and a wide difference is called a *weak basis.* A weak basis (cash far below futures price) reflects an oversupply and/or low demand situation, and a strong basis (cash close to or over futures price) reflects an undersupply and/or heavy demand situation.

The terms *being short the basis* or *being long the basis* are often used. If a merchant is long futures and short cash, he is short the basis because if the basis advances he will theoretically lose money, while if it declines he will profit. Conversely, if one is short futures and long cash, he is long the basis because if cash advances he will make a profit, while if it declines he will lose.

Cash and futures prices tend to converge as the futures contract expires, and thus the basis approaches zero, because the supply and demand factors creating the price of both the cash commodity and the expiring future theoretically become identical at the time of actual delivery. If the nearby future were at a *discount* to (lower than) cash on the last trading day, for example, a trader would take delivery on the future and sell the commodity in the cash market on the same day and profit by the transaction; but since everyone would do this, the cash price would be depressed to meet the futures price. In actual practice, however, the costs of making or taking delivery need to be taken into account, and this generally causes a difference between the cash and futures prices, even on the last trading day of an expiring future.

If the basis is the same at the time the hedge is lifted as it was when the hedge was placed, the hedge is said to be a *perfect hedge.* Such perfect hedges occur only rarely in actual practice. Any change in the basis between the time the hedge was placed and the time it was lifted will yield a profit or loss, depending upon what type of hedge was placed. The major benefit of futures hedging is that a commercial firm can replace a major risk—involving the entire cost or price of the commodity in which it deals—with the much smaller risk involved in basis change. And with that risk also comes the possibility of additional profit if the basis moves in the hedger's favor. While hedging substantially lessens the risk of ownership, movements of the basis can have a significant effect on the profit or loss of a hedge because the basis is the one part of the transaction that cannot be hedged. Because the basis can change, hedging does contain an element of risk.

Basis Trading

In many commodities industries, especially in the grain trade, cash transactions are often made with no reference to flat price levels but only to the basis, i.e., the difference between the cash price and the futures price. In these industries, most cash prices are calculated from the futures prices, whether or not the actual cash bids or offers are expressed as flat prices or as basis bids or offers. For example, an elevator operator who is a flat price trader might offer his corn at $2.65 per bushel if the March future is at $2.69 and the basis is 4 cents under the March. If the futures price changed, his offer would need revision. The basis trader, however, would express his offer as "4 under the March," and he would not have to change his offer until his basis changed.

In addition to stabilizing cash bids and offers, basis trading also allows cash merchants to make forward sales and purchases at a certain basis and to establish the flat price at the time the commodity is exchanged. For example, an elevator operator may agree in April to supply corn during the month of June at 2 cents over the July future. He would offset this sale with a later purchase of corn at perhaps 1 cent under the July future (which would result in a profit of 3 cents per bushel). Because futures positions are not necessarily involved in such a transaction, the offsetting purchase of the cash grain must be completed as close as possible to his delivery on his short cash sale in order to avoid fluctuations in the July futures price. This transaction would be equivalent to agreeing to supply corn at $2.70 and covering this short sale with a purchase at $2.67.

Since the basis reflects the difference between the cash price and the futures price, the basis may remain unchanged while the actual price fluctuates over a wide range, depending on the fluctuations of the futures price. For example, if a buyer of spot (cash) corn is paying 2 cents over December for No. 2 Yellow corn and if December corn futures are $2.50, this buyer would pay $2.52 for No. 2 Yellow corn. If December corn futures then declined to $2.49, his cash price would be only $2.51, and if the futures then rallied to $2.51, the cash price would increase to $2.53, with the basis remaining 2 cents over all the while. Likewise, if this merchant were buying No. 2 Yellow corn at 2 cents under December and December corn futures were $2.50, cash corn would be worth $2.48. If December corn futures then declined to $2.49, cash corn would be worth $2.47, and if the futures rallied to $2.51, cash corn would be worth $2.49.

Types of Hedges

The two basic types of hedges are the *short,* or selling, hedge, and the *long,* or buying, hedge. The selling hedge is used by the owner or producer of a commodity to protect the price at which he will sell it. The buying hedge is used by a merchandiser, processor, or user of a commodity to protect the cost of the commodity he will buy.

A. The Selling Hedge

The selling hedge, or short hedge, involves selling futures contracts equivalent to the amount of the commodity to be hedged. By doing so, the short hedger has protected his price objective against a price decline. A short hedger is a commercial interest who has a short position (has sold) in the futures market against his long position in cash grain or grain products. Examples of this include a miller who buys wheat today that will not be converted into flour until next month, a terminal elevator that buys corn today that will not be sold for three months, and an exporter who buys oats today that will not be sold overseas until later this month. All of these interests may shift part of their risk of grain ownership for varying periods of time by selling hedges in the futures market.

To illustrate, assume a farmer expects to harvest 10,000 bushels of corn. Regardless of how he may sell it, he naturally wants to get a price that covers his costs and that returns a reasonable profit. If the farmer decides that a price of $2.40 per bushel will be adequate, he can try to insure his price by hedging in the futures market.

In May, the farmer looks at the futures price for corn to be delivered in Chicago in September. If his normal basis for September is 8 cents under the Chicago price, and if September corn is trading at $2.48 per bushel on May 20, the farmer may decide to hedge his $2.40 price in the September future on that day. To do so, he sells two contracts (10,000 bushels) of wheat. The farmer's position looks like this:

cash	**futures**
May 20 growing 10,000 bushels of corn on which price objective is $2.40 per bushel	sells 2 contracts of September corn @ $2.48 per bushel

Having placed the hedge, the farmer has several options for completing his hedge depending upon the movement of the September futures price in relation to the movement of the cash price of corn. If the cash and futures prices move in parallel fashion, with the basis either unchanged or strengthened by the time the farmer is ready to market his corn, he may sell it through the normal local channels and offset his futures position. If prices have dropped, he will have realized a profit on his short futures position that will balance his loss on his cash corn. If prices have risen, his gain on the selling price of his corn in the cash market will be balanced by a loss on his futures position. In each case, he will have realized or exceeded his price objective.

Assume, for this example, that both cash and futures prices have dropped by the same amount—that the farmer's basis has remained 8 cents under (a perfect hedge). In late August, when the farmer is ready to sell his corn, the best price he can get for it is $2.37 per bushel. This is 3 cents below his price objective, and it cuts into his profit margin. The futures position has been profitable, however, because the farmer can offset his short position with a profit of 3 cents per bushel. The profit on the transaction yields a net price of $2.40 per bushel, the price for which the farmer had hedged. Because he hedged, the farmer received the profit he had decided he needed. The accounting would look like this:

cash	futures
May 20 growing 10,000 bushels of corn on which price objective is $2.40 per bushel	sells 2 contracts of September corn @ $2.48 per bushel
Aug 20 sells 10,000 bushels of corn @ $2.37 per bushel	buys 2 contracts of September corn @ $2.45 per bushel
result: $0.03 less than price objective	result: $0.03 profit

cash price received for corn	$2.37 per bushel
gain on futures	+ .03 per bushel
net price received	$2.40 per bushel

The farmer also has the option, of course, of delivering his corn in satisfaction of his short futures position if the corn is of contract quality and if the futures price is higher than cash prices. The costs of arranging to deliver in Chicago, however, will often dictate against the short hedger doing so.

If the movement of the cash and futures prices has not been parallel, and the basis has widened, offsetting his futures position and selling his cash corn will not provide the farmer with an adequate price for his corn. In that case, the farmer has the option of rolling his hedge forward, i.e., offsetting his short futures position in the one delivery month and initiating a similar short position in a futures month that offers better potential for protecting his price objective. The farmer would then hold his corn until his basis improved.

Short hedgers would not have too many problems if cash grain were always priced at a sufficient discount to the futures and hedged (as in the example above), because this would permit them to make delivery against their short sale and capture the difference between the buying basis and the delivery basis. However, a premium basis situation, where cash grain costs more than the nearest futures price, sometimes occurs. When grain is bought at a premium to the nearest futures price and a hedge is sold, the prospect of delivery against the short futures position must be ruled out, and merchandising the grain becomes virtually the only possibility for a profit. The cash price and that of the nearest futures tend to converge as the expiration of the particular futures month draws near, working against the short hedger who is long cash at a premium to the futures.

B. The Buying Hedge

The buying hedge, or long hedge, involves the purchase of futures contracts to protect the cost of a commodity not yet acquired against possible price increase. The buying hedge can be used by all who purchase grains and grain products, such as elevators, feedlots, processors, manufacturers, exporters, and end users. It is especially valuable for those who enter into forward contracts with their customers because it provides an insurance mechanism for the cost of the commodities that will have to be purchased to fulfill the forward contracts.

For example, flour millers often buy wheat futures against forward sales of flour. Although the commercial baking industry prefers to buy flour at a set price for delivery over long periods of time, millers do not stock nearly enough inventory to provide for this need, and they buy large numbers of futures contracts to figure their best price on flour to the bakers. Exporters are also users of long hedges against sales of grain overseas. When foreign buyers ask for offers on U.S. grain, exporters very often don't own the necessary inventory, but they confidently make offers, knowing that if they are accepted they may protect themselves against much of the risk of cash prices advancing by buying an equivalent quantity of futures and then offsetting that purchase of futures at the time the purchase of cash grain is made.

As an example of a buying hedge, assume that in January an exporter receives an order for 300 tons of soybean meal to be shipped in six months at a certain price. The price the exporter sets for this order is based on the cash price of soybean meal on the day he makes the sale, although he will not buy the meal until he is ready to make delivery. To protect himself against an increase in the price of soybean meal between the time he has sold the meal and the time he will buy it, the exporter hedges in soybean meal futures contracts. Assume that he has sold the meal based on cash meal costing $125 per ton in January. He will buy three contracts (300 tons) of soybean meal for July delivery at the time he makes the sale. Assume that July futures on that day are quoted at $128 per ton. The accounting would look like this:

cash	**futures**
Jan 10 *sells 300 tons of soybean meal* @ $125 per ton	buys 3 contracts July soybean meal @ $128 per ton

During the life of the hedge, the possibilities for price change are similar to those for the short hedger, except that for the long hedger a weakening basis points toward additional profit. The long hedger may choose to take delivery on his futures position if delivery in the Midwest is economical for him. Most long hedges are offset before the futures contract matures, however, just as most short hedges are, because the futures market is being used for price protection rather than as an alternate market for the cash commodity.

Assume that when the exporter goes to buy the meal to fill his short sale, both cash and futures prices for soybean meal have risen $1.00 per ton. He will have to pay more for the meal he buys, but the profit he makes on his futures allows his net cost for the meal to be the price on which he based his sale. The accounting would look like this:

cash	futures
Jan 10 sells 300 tons @ $125 per ton	buys 3 contracts of July soybean meal @ $128 per ton
Jun 30 buys 300 tons of soybean meal @ $126 per ton	sells 3 contracts of soybean meal @ $129 per ton
result: $1.00 more than cost objective	result: $1.00 profit

cash cost of soybean meal	$126 per ton
gain on futures	- 1 per ton
net cost	$125 per ton

Out-of-position Hedging

It is not necessary that a commercial firm be able to make or take delivery on the futures market in order to use the futures effectively for hedging. When grain is located at a point where it cannot reasonably be moved into the terminal market specified for delivery against a short futures position, it is said to be *out of position.* It cannot be delivered in fulfillment of a futures sale either because of contract specifications or location and distance. Although the grain trade is tied together by a price-making structure and a domestic freight-rate structure that generally equalize the basis at various markets, in the final analysis futures prices reflect supply and demand conditions in the futures market delivery area. Over the short term, a set of economic conditions could develop in that market area, which might distort the normal price relationship between that location and the location of the out-of-position grain. This possibility must be closely watched by out-of-position hedgers.

Carrying Charges: Relationship between Futures Prices

In addition to the basis, the successful hedger must also consider the relationship that exists between the prices for different futures delivery months for the same commodity—both the direction and the extent of the difference.

Adequate supplies of a commodity usually cause the more distant futures to be at a premium to the nearby futures; this is referred to as a *normal market.* In such a market situation, the limit of the difference, or *spread,* between the futures months for a stored commodity is defined by the actual costs of carrying (owning) the commodity over time—storage, interest, and insurance. This limit is referred to as the *full carrying charge,* and it varies among different commodities. (See Chapter 3 for a discussion of the calculation of carrying charges.)

Except in extremely bearish markets, full carrying charges are rarely achieved in actual market behavior. In the actual market, the spread between the months varies considerably, depending upon the supply situation and the relationship of cash to futures prices. Consequently, carrying charges, as reflected by the differences in price between contract months, are only indicative of an average situation that exists in the marketplace at a particular moment, and they are a function of the hopes, fears, and biases of all traders in regard to demand for those futures months.

If we assume that full carrying charges for soybeans are 6 cents per bushel per month, for example, the prices in a normal (but not full carrying charge) market for soybeans might look like this:

September	November	January	March	May	July	August
$5.50	5.61	$5.72	$5.83	$5.94	$6.05	$6.16

When the spread approaches or is at full carrying charges, the premium contract month offers excellent hedging opportunities for producers or owners of the commodity; and it may even be beneficial to lift a short hedge in a discounted month and place it in the month with the larger premium. This is called *rolling the hedge,* or *switching.*

Where the situation is reversed, when supply and demand indicate that a shortage exists, the premium will narrow on the deferred contract months. If a scarcity develops, the carrying charges will disappear or actually invert. This situation is called an *inverted market,* and it reflects negative carrying charges. Scarcity causes high prices in the cash and nearby futures contracts because the market gives priority to the present and discounts the future. The market says to the holder or producer, "We will pay you a premium for your commodity if you deliver it to us now." Futures prices in such a market for soybeans might look like this:

September	November	January	March	May	July	August
$6.04	$5.95	$5.86	$5.77	$5.68	$5.59	$5.50

In such a market, cash prices might well be over the futures, with the specific basis dependent upon local supply and transportation factors. In an inverted market, the maximum extent of the spread between the futures months cannot be defined; a discounted future can sell at any price below cash or another future.

A normal inversion occurs between the last futures delivery month of one crop year and the first delivery month of the next crop year because of the change in the supply situation—from the last of the previous year's crop to the harvest of the current year. For example, since October is the principal harvest month for corn, the December corn future is the first futures month of the new crop year, and December corn prices are often below those of the September, which is the preceding futures month. The size of this inversion is, of course, dependent upon the extent of carry-over from the previous year's crop and the anticipated size of the current harvest.

Consideration of the direction of the market, whether normal or inverted, and the extent of the carrying charge or inversion are especially important for the hedger who is involved in producing and/or storing commodities. A hedge is usually placed in the delivery month closest to the time the hedger will buy or sell the cash commodity. A hedge is sometimes placed in the current or closest futures delivery month, since the nearby futures represents the immediate cash situation and the liquidity of the nearby month may be greater than a more deferred month. However, hedgers will normally analyze the potential advantages to be found in placing hedges in any of the available months.

Whether the futures market tends toward carrying charges or toward inversion affects a hedging program. A short hedger prefers to work in a carrying-charge situation because he can cover at least a portion of the cost of carrying his inventory and sometimes nearly all by moving the hedges forward. This is accomplished by buying in the short position and selling a later delivery month. The *spread,* (difference) between the price at which the hedge is bought and the price of the sale in a deferred month represents the part of the carrying charge that can be obtained. In an inverted market, however, the short hedger would have to pay more to buy in his short hedge than he could obtain by selling a later month. In that situation no contribution is made to the cost of carrying inventory, but rather a part of past carrying charge earnings must be forfeited for the protection of being hedged.

Exchange of Futures for Cash

In making a cash market transaction in grains, price is very often agreed upon in terms of the *basis,* that is, so many cents under or over the futures price, rather than expressed as a flat price in dollars and cents per bushel. Once the basis is agreed to, the buyer and the seller will execute their cash transaction, and liquidate their respective long and short futures positions in a single process. This process satisfies both their hedging and merchandising needs and establishes the agreed-to basis price.

For example, suppose a grain exporter commits himself to a forward sale of grain and hedges with a simultaneous purchase of futures. On the day when he must allocate the actual grain for shipment, he makes a bid to an elevator firm, expressed in basis. We will assume that the elevator has also hedged but with a short futures position to protect against a decline in the value of his stored grain. Both the buyer and the seller agree to the basis price. The short hedger—the seller—delivers the actual cash grain to the buyer, in return for the buyer's long hedge position in futures. That long position offsets the seller's initial short position, ending his hedge. In the process, the buyer—the long hedger—acquires the actual grain he needs for shipment and has assumed the short futures position of the seller, which automatically offsets his initial long position in the futures. Payment is made on the agreed-upon basis.

The reasoning behind the practice and recognition of transactions involving the exchange of futures for cash is that if hedgers were required to liquidate their futures positions in open outcry auction in the pit, they might have to do so at different prices. This would mean that a basis trade for the cash commodity would actually have been made at one basis for the long hedger and at a different basis for the short hedger; the effectiveness of the basis trade would have been lost. It is for that reason that many cash commodity dealers who trade on the basis often specify when trading the cash that they will do so only against an exchange of futures. If a merchant on one side of a basis trade requires an exchange of futures and the merchant on the other side is not hedged, the unhedged merchant assumes a net long or short futures position as a result of the versus cash transaction.

Conclusion

Hedging does not eliminate all risks due to price fluctuations, but it does limit the risk to that of a difference between the cash price and the futures price of a commodity. Such fluctuations are far less volatile and more predictable than fluctuations in either cash or futures prices. In addition, hedging cannot be, nor is it intended to be, a foolproof method for accruing additional profits. On balance, however, a prudent hedging program, assembled and implemented on a consistent basis, will be successful in minimizing the risk of price change and maximizing the opportunities for profit stability.

5 grading of grain

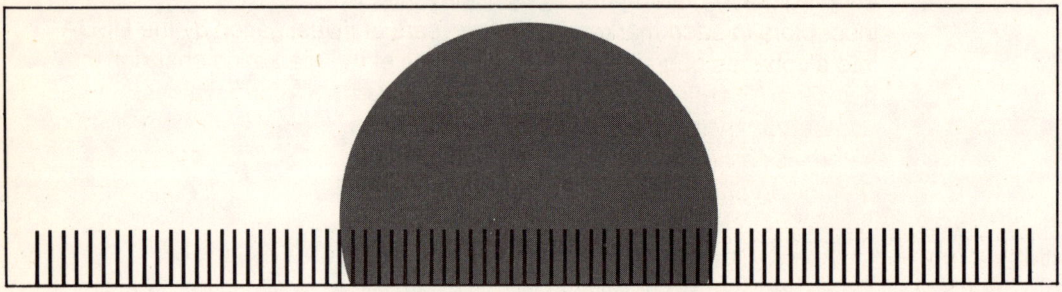

The federal grain-grading system, which is now used to classify and grade all U.S. grain, traces its origins to local and state grading systems. The Chicago Board of Trade, shortly after its formation in 1848, developed a series of grain grades and established a department to perform grain inspections. In 1871, the state of Illinois passed a law providing for grain grading by a state government agency, with the grades being revised at the beginning of each crop year. Soon other states began standardizing grain grading, although in some markets grading practices continued to follow local market standards only.

In 1908, the Grain Dealers' National Association established a set of uniform grades and inspections in the large grain markets. Though the association was not vested with enforcement authority, many markets adopted these grades and allowed their interpretation to rest with local inspectors in each market. Following years of investigation by the USDA into all phases of grain quality at all points of the marketing channel, the U.S. Grain Standards Act of 1916 offered the first uniform national standards. The act has been amended on a number of occasions to bring its provisions more into line with modern practices of grain storage, grading, and transportation.

Classification and Numerical Grades

Under the Grain Standards Act, specific grain grades have been established, which permit reasonably accurate determinations of quality consistent with the practical problems involved in the grading. The system is broken down into classes and subclasses determined by the shape, texture, and color of the kernel, and, sometimes, by the area of production. There are five classes for wheat, five classes for soybeans, four classes for grain sorghums, three classes each for corn and barley, and one class each for rye and flaxseed. For grading purposes, oats are no longer divided into classes, but rather into five carefully defined grades.

The act provides for a schedule of numerical grades based on quality, condition, and characteristics within the class or subclass. The highest numerical grade is No. 1, while the lowest is either No. 3, 4, or 5, depending on the grain and class. For some grains, separate grading schedules have been established for each class, while for others a single set of standards is provided for all classes. The numerical grade reflects the lowest grading factor. For example, corn at harvesttime often will grade No. 1 on all factors except moisture, which may be No. 5, thereby causing the corn to grade No. 5. The designation "sample grade," provided for all grains, means that one or more grading factors is lower than the minimum requirement for any numerical grade.

The most important factors used to determine the numerical grade of most samples are the following:

1. Test weight per bushel
2. Moisture content
3. Percentage of damaged kernels
4. Foreign material and, in some cases, broken kernels
5. Conditions such as rot, heating, sourness, mustiness; commercially objectionable foreign odor (COFO); and the presence of stones, cinders, or unknown substances

Test weight is a volume measurement to determine the weight per bushel and is a gauge of the plumpness of the kernel. It thus serves as an indication of the amount of product that can be recovered from processing. Since grains have different specific gravities, test weights of respective grains vary, just as test weights of the same grain may vary due to differences in environment.

The moisture content of grain is important because excessive moisture makes grain difficult to store and because buyers want grain rather than water. Moisture is a grading factor for corn, soybeans, and grain sorghums, and is measured to the nearest 0.1%. For wheat, oats, barley, and rye, excessive moisture causes the sample to be graded as tough, but it does not affect the numerical grade.

The percentage of damaged kernels is also a significant grading factor because damage may cause losses in processing. Damage may be due to fermentation, frost, fungus, sprouting, weather, fire, insects, or mold.

Foreign material is any extraneous material in the sample, such as dirt, weed seeds, other grains, or broken kernels of the grain itself; it is generally difficult to remove. The same type of material may be included in what is known as *dockage,* although dockage is usually more readily removed by machine or hand screening. Dockage is not a grading factor, but is designated as a percentage of the sample and is deducted from the total weight of the load.

Sample grade is assigned to grain having conditions of low quality, such as rot, heating, sourness, and mustiness, to samples having commercially objectionable foreign order (COFO), and to samples containing unknown substances, cinders, or stones above certain levels.

Other special designations are: *garlicky* (containing onions or wild garlic bulblets or pieces); *smutty* (containing smut balls or spores); and *ergoty* (grain that has been attacked by a fungus). These latter designations are noted on the grade certificate but do not ordinarily influence the numerical grade.

Most cash grain contracts provide for a contract grade—normally No. 2, but No. 1 for soybeans. In many cases the lower grades can be applied to the contract at the prevailing market discounts at the time the grain is shipped, or at a discount schedule agreed upon at the time the contract is made. Special designations also have a standard discount schedule.

Grain Inspections

Under the provisions of the U.S. Grain Standards Act, grain moving into intrastate commerce is no longer required to be officially inspected, i.e., sampled according to the act's provisions and inspected by inspectors holding USDA licenses. However, it is common practice for intrastate as well as interstate shipments to be officially inspected. Official inspections are performed by state agencies, by some commodity exchanges, and by private inspectors who make inspections themselves or through local chambers of commerce. All storage and transportation is done under the eyes of fully licensed inspectors and samplers. In the Chicago area, grain inspections are handled by an independent inspection organization. These men are licensed and bonded by federal and state authorities.

The country elevator settles with the farmer on the basis of federal standards although official inspections usually are not available. Beyond this point in the marketing channel, however, most shipments are officially inspected. Licensed inspectors are available in approximately 160 inspection points in the U.S., concentrated in grain-producing areas and at primary markets.

A. Obtaining the Sample

Upon arrival at an inspection point, boxcars containing grain are switched by the railroad to an inspection track in the inbound railroad's yard. Railroad personnel *bulletin* (list) the car numbers of these cars each day for the inspection service. After checking the number on the boxcar against the bulletin, the inspection service's sampler breaks the seal on the boxcar's door. (The seal, a thin metal strip imprinted with a number, is placed on the boxcar's door by the shipper and should be unbroken at destination, indicating that the car has not been entered since loading.) Then the sampler opens the car door and climbs up into the car carrying a sample bag and a trier—a long, hollow, slotted, double-shelled metal tube. The trier is inserted at a 10° angle into the grain while the slots are closed. When the trier is properly positioned, the slots are opened by twisting the handle, and grain from all levels of the car falls into the trier. It is then twisted closed and pulled out, and its contents are dumped on a sample cloth. Five individual probes from different areas in the boxcar are laid out on the sample cloth and examined for reasonable uniformity. If the grain is uniform, it is dumped into the sample bag. If the individual probes show a difference in quality, then more probes will be taken to obtain a fair representation of the contents of the car. The sampler then leaves the boxcar, closes the boxcar door, and places a new seal on it. An identification tag is placed in the sample bag showing the sampler's name, the boxcar number and its location, and any unusual conditions found in the car, such as an uneven load. The sample is then sent to the inspection service office for testing. Outbound shipments from subterminals and terminal elevators are probed and tested according to the same procedures as described for inbound cars.

Most inbound truck shipments are sampled by an automatic device that takes grain off the receiving belt in the elevator. Each truck lot is assigned a number and an official inspection is done for each. Outbound truck shipments are rare, but when they do occur they are usually submitted to inspection.

Inbound grain on barges and boats is usually sampled in the elevator by taking grain off the receiving belt at certain intervals, thus insuring a representative sample. At times, the sample may be obtained by probing with a trier. Samples of grain loaded out onto barges or boats are obtained by means of a mechanical sampler—a rotating scoop that takes periodic samples as the grain passes into the vessel.

B. Testing the Sample

Official samples are taken to the inspection service laboratory. All sample tests are made by qualified inspectors licensed by the USDA. Inspection procedures and equipment at all official inspection laboratories are standardized for each of the grains. After the sample has been thoroughly inspected, an official certificate is issued to the interested parties showing the inspection service performing the examination, the location of the grain examined, the car or barge number or boat's name, and the numerical grade showing all the factors. The grade shown on the certificate serves as one of the bases of settlement between buyer and seller. The shipper or seller customarily pays the cost of the inspection.

C. Appeal Inspections

In a small number of cases, a difference of opinion arises on an official inspection of a particular lot of grain. Either the buyer or seller may appeal the inspection as follows:

1. Call for a reinspection.

2. Request a federal inspection—sampling, analysis, and certification by the USDA District Grain Office.

3. If the grade is still in dispute, then request a Board of Appeal—review and analysis by the chairman and members of the Board of Appeals and Review, USDA Grain Division.

4. Appeal to the Secretary of Agriculture.

Appeals for inspections by Board and appeals to the Secretary of Agriculture are rare, since the parties generally accept the federal inspection. Each higher appeal rescinds the previous inspection. All appeal inspections are made by licensed inspectors in the employ of the federal government (who also supervise the work of licensed inspectors and their samplers). Appeals are available for grain shipped by truck but the load must be left intact on the truck for appeal sample. Practical considerations of truck management limit the number of truck appeals called.

6 transportation of grain

In its most basic sense, transportation is one of the means used by civilized man to bring order out of chaos. Transportation extends into every phase and facet of our existence. When viewed from the economic, political, and military standpoints, it is unquestionably the most important industry in the world. The more complex life becomes, the more indispensable are the things that make up a modern transportation system.

Transportation is very much an integral part of the grain-marketing process in that it is the vital link connecting the area of production with the area of consumption. Transportation costs are, therefore, important factors in the distribution of grain. The improvements in the inland waterways system and the construction of interstate highways have stimulated keen competition for the transportation of grain among U.S. rail, water, and motor carriers.

Grain is moved from surplus-production areas to deficit-production areas and from the areas of production to the points of consumption, milling or processing, and export. Theoretically, the differential between grain prices at any two points represents the cost of transportation between them. Price differentials greater than the shipping cost will cause a movement to a grain-short area in sufficient volume to narrow the price differential to the shipping cost. Conversely, a price differential less than the cost of shipping stops the movement to the deficit area until shipping costs can be recovered.

Changes in transportation costs can cause far-reaching economic effects not only in the surplus and deficit areas themselves, but also in relationships among terminal domestic and export markets, transportation modes, individual carriers, and the location of grain elevators and processing or milling plants.

An increase in transportation costs usually results in a combination of lower prices paid producers in the surplus area and higher prices paid by consumers in the deficit area, coupled with a decrease in transportation usage. The opposite prevails when transportation costs are reduced.

Each mode of transportation has certain unique advantages. Railroads and barges are generally considered the long-haul modes, while trucks specialize in shorter hauls. Intermodal competition is today's liveliest and most controversial transportation topic.

Historical Data

In the late 1830's, railroads became the leading transportation mode. The economic importance of this is unquestioned, but early rail history is replete with abuses to the shipping public in the form of discriminatory rates. In transportation terminology, *discrimination* refers to the granting of an advantage to one shipper, locality, or commodity that is not granted to all under similar transportation circumstances and conditions.

Efforts to pass federal legislation to correct these abuses were defeated several times. Finally, the original Act to Regulate Commerce was enacted February 4, 1887, under the authority of the Commerce Clause (Par. 3, Sec. 8, Article of the Constitution), and it is now called the Interstate Commerce Act. This act was a formulation of the common law and provided that all rates must be "reasonable and proper." Additionally, it provided for the establishment of the Interstate Commerce Commission (ICC) to be constituted as a quasi-judicial and quasi-legislative arm of Congress to administer, interpret, and apply the provisions of the act.

Unfortunately, the act failed to define "reasonable and proper" rates, and shippers and railroads were able to continue evading the intent of the legislation, which resulted in several Supreme Court decisions and numerous Congressional amendments.

Over the years, the ICC has assumed a constantly expanding role in the management of railroad affairs, especially rate-making. Competition from water and truck carriers was largely ignored by the ICC until the depression of the 1930's forced many rail companies into bankruptcy or receivership.

The early development of the trucking industry in many respects parallels that of the rails. Because of relative ease of entry and small capital requirements, the trucking industry experienced ruinous competition within itself and with other modes. To correct these abuses, Congress passed the Motor Carrier Act of 1935 as an amendment to the original Act to Regulate Commerce. Later, the Transportation Act of 1940 brought water carriers under federal regulation. Rail grain rates have been under the jurisdiction of the ICC since its inception; however, truck and water grain rates are specifically exempted from coverage by the amendments to the act.

Railroads were the principal movers of grain from the mid-1800's through World War II. Since World War II, however, and particularly in recent years, the railroads have been faced with rising costs, necessitating an increase in their rates to a point where grain shippers have sought other modes of transport. Each rail-rate increase has sent more traffic to competing trucks and barges, resulting in a substantial decline in the rail carriers' share of the grain traffic. Improved highways and water routes have expanded the areas accessible to trucks and barges, reduced their costs, and made them strong competitors. The interstate highway system, improvements of navigation on the inland waterways, and the St. Lawrence Seaway are examples of this trend. The trucks and barges, offering bare point-to-point transportation services, have had a lower general cost structure than the railroads, whose rates have included added costs for circuitous routings and other accessorial services. These services provide flexibility in grain merchandising and commonly include stopping for transit; the elevating of grain at industrial facilities; or stopping to reconsign grain; stops for inspection; stops for diversion of grain to a different consignee or location; and special free time, a guarantee, for instance, that grain will be unloaded within 24 hours rather than 48. Hence, competitors of the railroads found it easy to establish rates lower than the railroads could offer in areas where they wanted additional traffic. The frequent shortages of boxcars, deteriorating rail service, and a generally complacent attitude on the part of the railroads arising from their nearly monopolistic position in transportation all compounded the situation.

In 1958, the railroads, in an attempt to reverse the trend to these competing modes, began a program of selective rate reductions in their grain rate structure. In some areas, these reductions were quite drastic and were successful in retaining traffic, and, in some instances, they generated additional tonnage.

Freight Rate Construction

Freight rates, or transportation costs, are a very significant, though frequently neglected and dimly understood, component of the price system of a highly developed economy.

The primary basis of all freight-rate construction is the class rate in which every known commodity is placed in a generic class established for articles with similar transportation characteristics. Light, bulky articles of high value are given the highest rating, Class 100. Lower-valued commodities having a greater density and easier handling characteristics are assigned ratings downward from Class 100. The actual class rate structure is constructed strictly on a mileage basis, figured for the shortest route or combination of routes over which traffic may be interchanged or moved. For example, the short-line rail mileage between Chicago and Indianapolis is 181 miles. The Class 100 rate for 181 miles is $4.22 per hundredweight (cwt.), and all commodities assigned a Class 100 rating would be carried in carload service at this rate. A Class 60 rate would be $2.54, or 60% of the Class 100 rate, and likewise the rates for other class ratings are assigned at respective percentages of the established first-class rate.

Rail Rate Structure for Grain

Transportation charges for grain moving by rail are set on a specially designed commodity rate structure. Within the last decade, significant modifications of this rate structure have occurred because of the influence of truck and water competition, the introduction and widespread use of multiple-car and unit-train rates for domestic and export movements, changes in costs of providing transportation through technological advance, and the pervasive competition of private ("do-it-yourself") carriers. The grain rate structure is still characterized by peculiarities having no parallel in the rates on other commodities, although recent years have witnessed a tendency to place more grain and grain-products rates on a rate structure very similar to the class rate structure.

Rail grain rates apply equally to all shippers and are filed with the regulatory bodies in schedules called *tariffs.* The rates are published in cents per cwt., except multiple-car and unit-train rates, which are shown in cents per net ton. Grain rates for truck and water movements may be published in tariffs but more often are a matter of individual negotiation.

The grain rate structure of the past predicated on the value-of-service concept in rate-making, which gives consideration to the economic conditions surrounding the production and sale of a particular commodity, taking into consideration what the service is worth to the shipper. Today, railroad rate-making is directed toward a commodity rate structure based on cost-related factors, which are generally measured by the use of specific cost formulas approved and adopted by the ICC. Rail Form A provides various tables to be used as guidelines by the rail carriers in determining variable and fully allocated costs. These cost tables were developed by the Commission from operating-cost data supplied by all Class I line haul railways (railroads having revenues of $5 million or more) and arranged according to specific regions. Separation by regions was established because of differences in costs for each.

Costs based upon the ICC formulas are frequently introduced in rate cases by the rail carriers to justify their proposed revisions in their rate structure. In computing costs for single-line application, a carrier's costs may be below the average costs established by the Commission, and thus the rates may be compensatory even though they do not measure up to the requirements spelled out in Rail Form A.

Some of the other significant factors taken into consideration by the rail carriers in making rates are:

Rates of competing carriers and modes
Rates on the various grains
Competition among producing areas and markets
Volume of traffic and periods of movement
Type of equipment needed
Whether the equipment will be returned loaded or empty
Distance of the haul
Whether or not the rate will increase the volume hauled

The traditional railroad grain rate structure was based on west to east movement of grain and tended to equalize broad producing and consuming areas in various ways. Shippers within certain area groupings received rates equal to those of markets within broad consuming area groupings, and thus many hauls of unequal length received equal rates. Grain could move at equal rates from various gathering points through terminal markets (such as Chicago) to areas of consumption even if the hauls were longer for some lots of grain than for others. Transit arrangements gave shippers the right to store, mix, blend, or process grain at many points between the producing and consuming areas for no extra charge and to move grain products at the same rates as grain. Since transit points were not always on a direct line between the origin and the destination, transit rules permitted circuitous movement via competitive rail carriers without additional charge.

This grain rate structure made possible a relatively simple pricing system at terminal grain markets, which were concentration points for large quantities of grain, centers for cash and futures trading of grain, and, therefore, logical pricing points. Prices in producing areas were based on terminal market prices less freight charges to the market. Risk was minimized as a factor in prices because of the flexibility in equalized transportation costs between broad areas and over various routes. The transit system thus minimized the effect of transportation costs on plant location. The railroads and their historic grain rate structures were the principal reasons for the development of terminal markets, which allowed the grain industry to stabilize and then mature. Rail economists now point out that the changing pattern of grain movement in this country requires a critical analysis of rate equalization and transit concepts. This has resulted in drastic changes in the historical rate-making practices.

To regain traffic lost to competing modes, the railroads have cut their rates on a selective basis. This was done quickly and sharply in areas dominated by truck and water competition. These reduced rates have restricted application and do not include all of the services traditionally provided by the rail carriers. The number of transit stops has been reduced or the stops have been eliminated, transit charges have been added or increased, routings restricted, minimum weights increased, and charges added for accessorial services. In many instances, the reduced rates were not confined only to multiple-car, unit-train, and volume shipments, but have been extended to single-car movements at higher minimum weights. The grain rate structure in its present form has been truly described as a "monster."

Rate-break System

The topography of the U.S. includes several natural barriers—such as mountains, rivers, and lakes—and land areas separated by barriers have been designated as transportation rate territories. For the purpose of making rates, the country is broken down into five major rate territories: Official, Southern Freight Association, Western Truck Line, Southwestern, and Trans-Continental. The map below illustrates these rate territories. (Three of these territories are subdivided.)

1 Official
- Central Territory
- Trunk Line Territory
- New England Territory

2 Southern Freight Assn.
- Carolina Territory
- Southwest Territory
- Mississippi Valley Territory
- Buffer Territory

3 Western Trunk Line
- Trans-Mississippi
- Northwest Territory

4 Southwestern

5 Transcontinental

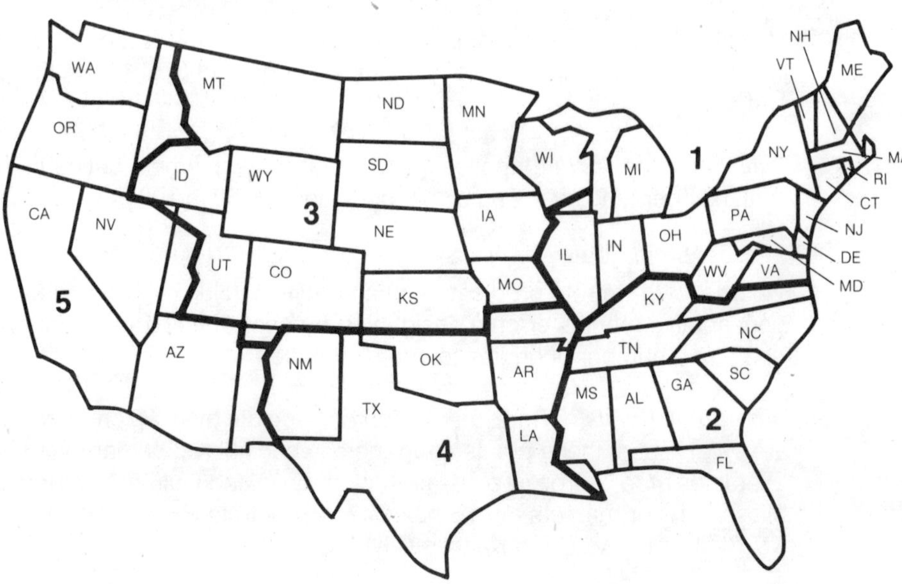

One of the foundations of rate making is to determine the proper transportation charge between two points and then allocate that portion of the total charge applicable to each rate area and to the individual carriers participating. For single-line movements, all revenues derived from a shipment accrue to the one railroad involved in the haul.

Terminology

The following are some of the terms or phrases necessary to understand grain transportation:

Flat or Gathering Rate—either a local rate of a single carrier or a joint rate of two or more carriers published as a unit and not dependent on any other rates.

Through Rate—a rate applicable to a given shipment from the point of origin to the final destination. If the through rate is applicable over the lines of several carriers, it is a joint through rate.

Proportional or Reshipping Rate—a local or joint rate dependent for application upon (1) previous transportation to the point from which the proportional rate applies; (2) a subsequent transportation from the point to which the proportional rate applies; or (3) both. It is supposed to be a part of a through rate and is usually lower than the flat rate between the same points. The oldest proportional rate published applies from Chicago to New York. All other rates in Official Territory are related to this, usually on a mileage percentage basis. The rate is important to the Chicago futures market because under regulations governing deliveries on futures, the warehousemen must guarantee that the owner of a warehouse grain receipt will pay a rail rate no higher than the published proportional rate from Chicago to New York.

Multiple-Car Rate—a rate applicable on several carloads moving as a unit between two stations. Generally, multiples consist of 3, 5, 10, or 25 carloads.

Trainload Rate—a rate applicable on a unit-train shipment between two stations operating as a single unit requiring from 50 to 100 covered hopper cars.

Volume Rate—a rate predicated on a specified volume over a period of 12 consecutive months or on a designated number of unit trains from one origin or a number of origins to one destination.

Rate-break Points or Rate-break Markets—points over which grain rates are established; these points are generally considered primary markets. Examples of rate-break points are Minneapolis and Duluth, MN, and the Missouri River markets from Sioux City, IA, south to Omaha, NB; Atchison, KS; St. Joseph, MO; and Kansas City, MO.

Final Markets—those markets located at the eastern terminals of the western railroads, generally considered the last storage points for western grain. They also act as rate-break points over which freight rates are made on shipments destined to the east and southeast. These markets include

Chicago, Peoria, and Cairo, IL; St. Louis, MO; Milwaukee, WI, and Memphis, TN.

Gateway—a point at which freight moving from one territory to another is interchanged between transportation lines. Frequently gateways are also rate-break points. Other gateway points include river crossings such as Evansville, IN; Louisville, KY; and Cincinnati, OH.

Equalization of Rates

Equalization of rates allows different areas to be equally competitive with respect to transportation charges. The grain rate structure was historically based on an equalization of rates on traffic moving from the west to the east at the time the railroads enjoyed a virtual monopoly in the movement of whole grains and their products. With the influx of competitive modes of transportation—water and truck—the railroads set out to meet this competition by making piecemeal adjustments in the total rate structure. The end result was a complete disruption of the rate structure. This disruption first surfaced with the feed grains. Since the carriers made no revisions in wheat and wheat-flour rates, this segment of the industry still had equalizations of the freight rates via the primary markets and rate-break points, until the recent rapid successive across-the-board freight rate increases eroded this position. Flour millers located at intermediate stations between the west and east no longer have equalization of freight rates because the general freight increases have completely disrupted the freight rate structure via the primary markets and rate-break points. Since wheat originating in the west and moving via the primary markets has numerous destination possibilities, its ultimate destination will be decided by the highest market price, including transportation costs, whether it be priced at the major primary markets or at intermediate wheat-milling stations. With an equalized freight rate structure, markets are equal insofar as transportation costs are concerned; however, the prices offered by the markets, taking into consideration the present-day transportation costs, will not be the same and the market with the lowest price offering would draw grain to the exclusion of other markets until these demands are satisfied at the premium to transportation costs, or the other markets meet or exceed that premium.

Grain pricing, then, reflects a constant adjustment of supply and demand inputs with the cost of transportation being an ever-present and literally fixed cost.

Transit Arrangements

Transit is the application of through-freight rates to a shipment stopped at a point intermediate to final destination for storage and/or milling and processing. This arrangement is accorded by a rail carrier to an industry located along its tracks upon application and approval and is extended to rail movements and certain combinations of haul between railroads and other modes. In addition to grain and grain products, transit is also granted to a wide variety of other commodities, such as steel, paper products, food products, and lumber.

Without a transit arrangement, the prices on many grain products would be substantially higher. Many attacks have been hurled at the transit concept, pointing out that it is merely a subsidy to certain commodities and therefore eliminates those commodities' share of the very high fixed-cost structure of the railroad industry.

To illustrate a transit application, assume an elevator company located at Point B buys grain from Point A for later sale to Point C. A flat rate of 30 cents per cwt. exists from Point A to B, and 50 cents per cwt. from B to C. The through rate from A to C via B is 70 cents per cwt. When the elevator operator buys grain from A and ships it to his elevator at B for storage and conditioning, he pays the railroad 30 cents per cwt. Some time later, because of a transit arrangement, he can ship the grain from B to C and pay only 40 cents per cwt., the transit balance of the through rate (70-30=40), instead of 50 cents per cwt., the flat rate applicable if a transit arrangement were not in effect. A similar arrangement could exist for grain moving into a processing point and the grain products moving out.

The railroads demand a record of all inbound and outbound shipments of commodities subject to transit applications. These records are audited by railroad personnel periodically to insure proper application of the granted arrangement. The life of rail-transit billing is generally two years, although this can be extended under certain conditions and upon payment of an extra charge.

Other Services

In addition to transit arrangements, railroads provide grain shippers with other services such as inspection-in-transit, reconsignment or diversion, weighing, free time for loading and unloading, out-of-line hauls, and switching. Depending on individual circumstances, some of these services are performed at a charge in addition to the line haul rates.

At the shipper's request, carloads of grain may be stopped for inspection for a grade determination in transit at a particular point en route or at destination. The shipper must pay for the actual sampling, inspection, and grading of the grain, since this is accomplished by an independent or third party and not under control of the rail carrier. With the introduction of automatic sampling devices installed at country elevator locations, the sampling and grade determination of grain is being accomplished in many areas at its origin.

Under certain conditions, the shipper may change the consignee and/or destination of the car while it is en route, at a junction-point gateway or carrier interchange point, or just prior to delivery at billed destination. Rail grain is often billed or ordered to an interim party or destination and then reconsigned to a final destination. In some cases, one change in the bill of lading is permitted without charge, although a charge is generally assessed for this service.

Free time refers to the maximum time allowed by a railroad to physically load and unload grain at origin and destination or to receive disposition orders on a shipment. Shippers and/or receivers who exceed the free time are subject to a penalty charge referred to as *demurrage.* These charges are graduated according to the length of time cars are held and are published in specific detention-charge tariffs.

New Developments

The pressures of competition can be a powerful motivating force in the improvement of productivity. The formidable motor and water competition facing the rail carriers has forced them to adopt innovations and new technologies in the movement of grain and grain products. The technological advancement of equipment has contributed immensely to the competitive forces in the transportation of grain. The development of covered hopper cars, with a capacity of up to 100 tons, proved beneficial to both railroads and shippers. To the rail carriers, it meant moving larger volumes at one time with greater ton-mile earnings and the ability to pass on some of these savings in the form of freight rate reductions. To the shippers/receivers capable of loading and unloading this equipment, it afforded wider marketing opportunities through reduced rates.

The motor-carrier industry has likewise come forth with redesigned tractors that allow the use of longer trailers. The interstate highway program, begun in 1956, has had a great influence on highway transport, particularly in the movement of grain from country origins to terminal markets. Recent legislation increasing the maximum weight limits on the interstate highway system has contributed to greater earnings per load for the trucking industry. During the chronic shortages of railroad cars facing the grain industry, trucks have been able to make deep inroads in moving a larger portion of the grain every year. Today's trucks are also moving outbound shipments of grain products and feed ingredients.

The water carriers have kept pace with the competitive situation with larger and more efficient barges, coupled with the use of more powerful tugs, or towboats, capable of moving a greater number of barges in one tow over the inland waterways system.

Freight Rate Changes

Carriers are concerned with the maintenance of a normal profit, reflecting the return on investment expected after a comparison of returns on alternate uses of funds. A carrier will adjust its level of rates in response to inflationary increases in its costs. Since the effects of inflation are general throughout the economic system, and since most costs are affected by inflation, the carrier tends to see horizontal rate adjustments applying generally to all commodities as an appropriate response to a general increase in costs.

The principle of horizontal increases is not new in transportation history. In the early 1920's, the rail carriers advanced their rates sharply and across the board during an economic recession, just in time to give encouragement to the application of the newer technologies in trucking and inland waterways. Likeways, in 1931 in the very depths of the great depression, when the country was in the midst of one of its greatest deflationary periods, the carriers again advanced rates across the board by substantial percentages.

When carriers advance rates upward, they do so in great sweeps with across-the-board increases; however, when freight rates are reduced, this is generally done in piecemeal fashion. At times, these downward adjustments have been accomplished while the carriers are pressing the regulatory bodies for general increases. This has been found to be true where adjustments were being made to meet a competitive situation while the carriers sought authority from the ICC for a general increase on all commodities. In recent general freight rate increases, the rail carriers on their own have allowed hold-downs, or exceptions, to apply on selective commodities. This was done to stablilize a competitive situation.

The grain rate structure has been in a constant state of flux. Some changes originated with shippers/receivers, while others were carrier proposals designed to meet competition, with the net effect being a rapidly changing rate structure that has greatly altered traditional marketing methods. Greater use and larger fleets of covered hopper cars have resulted in the introduction of unit-train and multiple-car rates, through which the railroads have been able to make substantial reductions in the cost of transporting grain and grain products by rail. The export movement offered the greatest potential for these reduced rates because of the larger volumes involved from large inland terminals to large export terminals. Many of these rate adjustments, directed solely to the large-volume shippers, resulted in a disadvantage to the small shippers whose facilities could not accommodate the loading of multiple cars. These small shippers have been forced to expand their facilities to accommodate the loading of multiple cars and are now able to take advantage of lower export rates. The ten-car rates, to a great extent, have put these small shippers back into the competitive picture.

The unit-train rates are now being directed more to domestic marketing channels, primarily in the movement of wheat, corn, and soybeans.

The railroads have also shifted to a rate-making structure designed to eliminate accessorial services, such as transit arrangements, diversion and reconsignment, inspection-in-transit, and a shortened free time for loading and unloading. Point-to-point mileage-based domestic rates have evolved from this theory of rate making. Another example is gathering rates into major terminal and processing centers, applicable only when shipper-furnished covered hopper cars are used. This relieves carriers of the obligation to furnish their own equipment.

While competition in its many forms is a major factor in bringing about rate changes, it is not the only consideration; other forces also stimulate change. New harvesting methods and changes in the marketing system have all had a direct effect on carrier rate-making policies.

The needs of the world markets have altered the export grain rate structure. Rate changes favoring one range of ports is immediately followed with an adjustment to another port range. For example, adjustments by the eastern railroads in their export rates to the North Atlantic ports from the Midwest (Illinois and Indiana) is usually followed by comparable rate adjustments by the railroads serving as the Gulf Ports. While the Great Lakes have been officially designated as our fourth seacoast, railroads serving lake ports have been reluctant to establish an export-rate structure to put these ports on a competitive basis with the year-round well-established tidewater ports. The Port of Chicago through the efforts of the Chicago Board of Trade has been successful in securing export-rate adjustments from Iowa and Illinois origins located on western railroads that reach the port in single line operation. The export grain rate structure to Chicago now includes rates on corn and soybeans applicable in single-car units as well as units of 25, 50, and 54 cars.

In the nation's expanding economy, the demand for freight services is expected to accelerate in the coming years. The grain trade will contribute to this demand and the different modes—rail, water, and highway—will be required to respond with substantial investments, with additional support, no doubt, coming from governmental sources. The grain trade needs and will demand a viable transportation network to meet the demands of the domestic and export markets. Intermodal competition will remain strong and we can look forward to the utilization of joint intermodal operations. The accent will be on dependable and reliable service through increased productivity. Existing and future transportation facilities will be taxed to the utmost efficiency.

7

country elevators

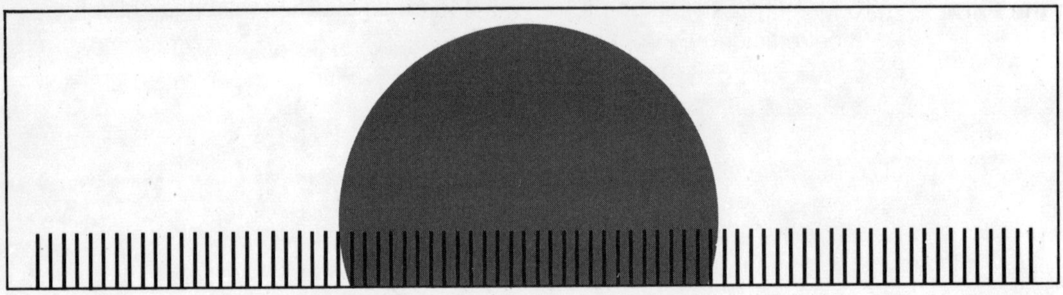

The marketing of grain by farmers in surplus grain-producing areas of the U.S. is the first step in a series of business transactions by which grain moves to processors and then to consumers. The farmer has a choice of several outlets for his grain, but the most common outlet is the country elevator located near the production area. The operators of country elevators buy grain from farmers under different plans and merchandise it in any of several ways. While the country elevator usually derives a considerable portion of its income from buying and selling grain, income is also obtained from conditioning grain, merchandising certain farm supplies, performing customer services, and providing storage for grain producers, processors, and the Commodity Credit Corporation.

Grain Movement from the Farm

The period of the year when farmers market their grain varies among grains, among years, and among sections of the country. Nearly half of the U.S. soybean crop is sold during the first three months of the crop year—October, November, and December. The peak of the movement of wheat, oats, rye, and barley is at their respective harvesttimes, when much grain is delivered directly from combines.

The practice of delivering corn to the elevator on the ear still prevails in some sections, but a more common practice is for farmers to hold corn in cribs on the farm to dry, and then to shell the corn on the farm before loading it on trucks for delivery to the elevator. Currently, most corn is combine harvested, i.e., shelled and harvested at the same time. Combining permits the corn to be shipped to the elevator directly from the field. Most of this corn is artificially dried before being stored in bins on farms or delivered to elevators.

While most grain is delivered at the harvest or shortly after, farmers are more and more frequently equipping themselves with storage facilities on the farm, making it possible to await a more favorable market, especially if prices at harvest seem depressed. A principal factor in this development has been the government-support program. On the farm, ears of corn are commonly stored in cribs designed to provide adequate ventilation for properly drying or curing the corn. In years of large crops, corn is also frequently stored in driveways, farm buildings, and temporary cribs made of fencing.

Storing small grains, especially wheat, on the farm presents more of a problem due to the difficulty in avoiding deterioration. In the storage of these grains, a bin above the corncrib or a separate steel bin is used. Storage bins are protected from rain, well supported to hold the weight imposed, sealed by sheathing, well ventilated, and as nearly vermin-proof as possible.

Farmers may use their own trucks or the services of a commercial trucking concern to move grain to market. Most grain is sold to a country elevator; however, the farmer may sell to an itinerant trucker who buys the grain on the farm and resells it wherever a suitable market is available, or to other farmers, usually livestock feeders who do not raise enough grain to feed their stock. At one time, farmers could sell grain directly to flour millers and other grain and soybean processors who had established plants in the grain-producing areas. Today, however, changing freight rates have encouraged the movement of processing plants back to consuming areas, although in some areas farmers are still able to deliver grain directly to the processor's plant. Farmers located near a terminal market can truck their grain to market where they can sell directly to terminal elevators equipped to unload trucks. Although these various outlets for grain are available, the great bulk of marketable surplus grain goes to the country elevator.

Types of Country Elevators

Country elevators fall into three general classes: (1) independent or privately owned elevators; (2) farmers' cooperative elevators; and (3) line elevators. *Independent elevators* are those owned and operated by individuals who provide their own capital and usually operate the elevators themselves.

Elevators owned by farmers include two types or organizations. The first is the strictly cooperative elevator established by an association of farmers, which has its own capital structure and returns profits to the members on a patronage dividend basis. The second type of farmer-owned elevator is composed as nearly as possible of farmer patrons and governed by a directorate elected from and by the stockholders with profits distributed as dividends on stock rather than on the basis of patronage. Commission houses in the central markets have often financed both private and farmer-owned country elevators, particularly in the Northwest, lending them money for the purpose of carrying on country buying operations.

Line elevators are chains of elevators owned by a grain company, usually running out along a railroad system. Many of the line-elevator companies obtained their start as a result of the practice of railroads of giving exclusive rights for the erection of elevators at specified distances along their lines. Line-elevator operators receive all instructions regarding buying prices and operating practices from the parent company, often a flour mill, grain processor, or grain import/export firm. In Canada, virtually all elevators are line elevators.

Whatever the organization setup of a country elevator, the services rendered are essentially the same, and the cost of these services varies little between elevators.

Facilities

A typical country elevator today has a minimum storage capacity of several thousand bushels to a couple of million bushels. It is no longer possible to distinguish by size country elevators from terminal elevators, which are located in the principal grain-marketing centers. Country elevators vary considerably in construction, but have many basic similarities.

A typical plant of 100,000-bushel capacity is constructed of poured concrete or of steel. A scale for weighing farmers' grain is placed in a driveway or other readily accessible position. Below floor level is a pit into which the grain is dumped from the truck by a mechanically operated truck lift. An elevator of this size would probably have from 12 to 15 bins with capacities of 1,500 bushels to 15,000 bushels each. Each bin may be equipped with an automatic loading-out scale with two outlets, one for loading railroad cars and the other for loading trucks.

Grain is moved to the different parts of the elevator by electrically operated machinery and by gravity. Endless belts are used to move the grain horizontally, and a leg, an endless belt carrying V-shaped metal buckets, raises the grain vertically. Spouts are used in the downward movement of grain by gravity.

Purchasing Farmers' Grain

General practices connected with the purchases of grain from farmers and its forwarding to central markets are fairly uniform among all types of elevator operators. The method of purchase varies with the different grains, the season of the year, the section of the country, and the policy of the individual elevator. Grain is obtained from farmers by outright purchases at the time of delivery, by contract made in advance of delivery, by accepting grain for storage with sale to be consummated at a later date, and by handling grain for the farmer's account without ever taking possession of it.

The method of purchase most widely used is outright purchase at the time grain is delivered. In order to insure obtaining a considerable volume of grain, however, elevator operators are arranging more and more advance contracts with farmers for delivery of grain, often for grain not yet harvested. Conditions stipulated in contracts vary under different circumstances, but usually cover the grade of grain, the time of delivery, and the price.

If a farmer believes that grain prices may move in his favor at a later date and he does not have storage facilities on his farm, he may make arrangements with the country elevator to store his grain. This practice has been used most in the storage of soybeans, wheat, and corn. The amount of grain that can be accepted for storage is limited because the elevator requires a considerable portion of its storage facilities for its own grain handling operations.

On rare occasions, when a farmer has a carload of grain that he desires to ship on his own responsibility, he may arrange to have his grain loaded through the local elevator into a car. In such cases, the farmer arranges all the details of shipment and the elevator simply charges an agreed handling charge, assuming no responsibility for the disposal of the product. It is very common for an elevator to provide for the transshipment of grain, making a nominal charge to the farmer for his service.

During the hours of trading on grain exchanges, the country elevator operator has access to continuous market information through commission merchants operating in the terminal markets. He also receives frequent quotations from radio reports on the grain markets and from interior dealers who purchase grains for resale to interior mills, processing plants, and terminal elevators. Many large merchandisers have begun to install their own ticker systems for price quotations. When the country elevator manager purchases a quantity of grain from a farmer at any hour of the day, he is simultaneously aware of the reselling opportunities for that grain; and he bases the price upon bids from terminal markets and interior dealers, taking into consideration the freight charges to the terminal market, the terminal market handling costs, and his own cost and profit.

Bids to the farmer are usually made for a specific grade of grain, with immediate discounts for lower grades sometimes specified at the time of purchase and other times determined when the grain reaches the terminal market. For example, when new corn begins to move in the fall, it usually carries excess moisture. For that reason new corn usually grades worse than No. 2. Country elevator operators bid for No. 2 with discounts to apply for factors that do not make grade; 15.5% is the maximum moisture allowable for the No. 2 grade. Usually, for each ½% moisture content, in excess of the permissible 15.5%, there is a discount of 1% in the value of the grain.

The specific method employed by the country dealer in determining the price to be paid to the farmer may be illustrated by the following example. Assume a farmer has corn to sell that grades No. 2 Yellow. The dealer's bid is based on the Chicago market, where No. 2 Yellow corn is quoted at $2.25 a bushel. The freight rate is 20 cents a hundred pounds, or 11¼ cents a bushel, on the basis of 56 pounds to a bushel of corn. The customary commission on handling corn in the terminal market is about 1 cent a bushel, and the country elevator usually requires approximately 4 cents for handling costs. Thus the price paid to the farmer must be at least 16¼ cents under the Chicago market price, or approximately $2.08¾ per bushel.

It is a relatively simple matter for the elevator operator to determine his bid price when he is situated near a terminal market where all grain bought is destined for shipment in the same direction. In territories where freight charges to all surrounding markets are nearly equal, however, it is necessary to keep a constant check on the different markets to determine the highest price. It is not unusual during the shipping season for grain from north-central Iowa to be shipped to Omaha, Chicago, Cedar Rapids, St. Joseph, Kansas City, St. Louis, Milwaukee, Minneapolis, and river ports or gulf ports for exporting. Market conditions at these different points will cause relative prices to change sufficiently to move grain in various directions. Also, it is often true that different markets may vary on relative prices of different grains, e.g., Milwaukee may be the high bidder on barley, Cedar Rapids on oats, and Chicago on corn.

Handling of Grain

Grain brought into an elevator by a farmer is first sampled and weighed in its vehicle and then dumped into the elevator pit. The sample taken is marked and brought to the elevator office for grading. The manager determines each of the grading factors.

After the grain has been unloaded, the empty vehicle is weighed, with the difference between the weight of the full load and the weight of the empty truck determining the quantity of grain delivered. When these operations are completed, the elevator operator usually gives the farmer a check for his grain, unless the farmer plans to make several more deliveries soon.

If the grain being delivered is of sufficient quantity and of uniform quality, it may be elevated directly from the dump to the scale hopper, through the automatic scale (if there is one), and directly through spouts to the car for shipment. The grain may also be elevated and distributed to a bin containing grain of a like quality, or if there is bin room and no grain of similar quality on hand, it can be distributed to an empty bin and retained there until sufficient grain of the same quality has been received to make up a carload.

When stored grain is loaded out from the bins, the grain flows from the bin, which has a hopper bottom, into the boot of the leg (the lowest part). It is then elevated to the distributor head, carried to the hopper scale, and loaded on the car or truck.

Merchandising Grain

The elevator operator must decide the best method for merchandising the grain stored in his elevator bins. In selecting the market for his grain, the elevator operator considers many factors, including the condition of the grain and the risk of quality deterioration while en route, the available transportation to market, financial conditions, the type of demand, the volume of movement, and the habits and customs of buyers. The main consideration determining the destination of grain, however, is price. A price difference as small as ¼ cent a bushel may determine the place of sale, although when the price is practically the same in all markets, the decision will depend on personal preferences.

The country elevator's grain can be sold locally or forwarded to a principal terminal market, to an interior mill or processing plant, or to an export point. When a local feeder is willing to pay a price exceeding the terminal market price (less freight, commission, and handling charges), the operator generally will sell to the local feeder, either loading the grain on the feeder's truck or delivering it to the feedlot.

Types of Sales by Elevator

The most common outlets for a country elevator's grain are the terminal markets and, increasingly, subterminals. Grain destined for terminal markets is either consigned to a commission merchant (cash merchant) or sold for deferred delivery on a to-arrive or on-track basis.

In consigning grain, the country elevator ships the grain to his representative in the market to be sold after its arrival. The country elevator retains ownership of consigned grain and is subject to any loss incurred while the grain is in transit, such as adverse price changes or damage to the grain. The seller pays all costs of shipping the grain to the market, not knowing what will be realized on the grain until after it has arrived at the market, been inspected, and offered for sale.

When grain is sold on a to-arrive shipment basis, the contract is usually made before the grain is shipped from the country elevator. In addition to price, the time of shipment, the grade, and the terms of sale are specified in the deferred delivery contract. The time specified may be 5, 10, 20, or 30 days, or in some instances as much as 120 days. Discounts for grain grading lower than the contract grade may also be provided. Inasmuch as the price is fixed at the time the contract is made, the country elevator is relieved of all risk incident to fluctuations in the price of grain, the buyer assuming these risks.

One type of deferred shipment is the on-track country elevator point sale (also identified as track country station, or TCS) in which the price is set f.o.b. country elevator point. In making such a sale the country elevator operator agrees to load the car within a specified period, and the buyer pays the cost of shipping the grain to the destination.

Another common type of deferred shipment is the to-arrive sale, in which the price is based on delivery at the destination point; the country elevator pays the freight costs in shipping the grain to this point.

Grain sold to arrive by the country elevator is usually handled in the terminal market in much the same manner as grain on consignment. The elevator's merchandiser or the commission merchant canvasses the market to obtain the best bid; he may accept this bid or apply the car on a previous sale. During trading hours on an exchange, bids fluctuate according to market changes, but overnight bids, because they are based on the previous day's closing prices on the exchanges, usually remain reasonably constant.

Some country dealers allow the purchases made after the close of the market to accumulate until the following morning when the dealer has not only the bids based on the previous close of the market but also bids from numerous brokers who desire to purchase grain for processing plants and others. The dealer may sell all his overnight purchases to one buyer or distribute them among a number of buyers, depending upon the price differentials.

The extent to which grain is consigned instead of sold for deferred delivery varies among different sections of the country, among different elevator operators, and according to the condition of the market and variations in the quality of grain received from farmers. The personal inclinations of the elevator operator are a most important element in this type of sale. If he feels the market is in a strong position, he will be more likely to consign and take the risk of a market decline while the grain is in transit; whereas if the elevator operator desires to obtain a known profit without further risk, he will sell to arrive or track country station even though he obtains less than the spot terminal market price, because in these cases the price is ordinarily a definite one, fixed before shipment is made. In recent years, to-arrive sales have become more widely used than consignment sales.

If the grain is of poor quality, it may be impossible to sell for deferred delivery, and consignment becomes the only alternative. During periods of heavy grain movement, however, the elevator man in the Chicago territory, for example, will probably sell at least a part of his purchases on a to-arrive basis. One reason for doing so would be to avoid the price decline on large amounts en route to market.

In some cases, the farmer offers the elevator his grain subject to a predetermined price or consigns his grain through his local elevator operator. In such cases the elevator man becomes little more than an agent and handler for the farmer.

In buying and selling grain, the country elevator is subject to the risk of loss from fluctuations in price, although part of this risk is eliminated when he sells grain to the terminal market at the same time he buys grain from the farmers. Sales on-track and to-arrive also provide a means of shifting the risk, since the price is fixed at the time the contract is made. There are, however, definite limitations to the extent to-arrive and on-track sales can be used to offset cash purchases. To-arrive prices do not necessarily stay in line with spot prices, and the elevator operator cannot always sell his grain by this method at a price that compares favorably with the cash price he must pay to his customers.

Use of the Futures Market

When a to-arrive market is unfavorable, if the elevator operator is to protect himself against price fluctuations, he must turn to hedging in the futures market. Country elevators use the futures market for hedging principally in the sale of futures against cash grain which is either in transit for sale upon arrival or held in store.

In those sections of the country such as North Dakota, South Dakota, Montana, and Minnesota where much time is required to ship grain to market and pronounced price changes may occur in the interim, the consignment method of selling is employed extensively. To overcome this difficulty, elevator operators in these areas frequently sell futures as cash grain is purchased and remove the hedge by a purchase of futures after the grain has been consigned to a market and sold.

Another use of the futures market employed by the country elevator operator is that of hedging grain held in storage to earn carrying charges when the difference between the futures contract price and the cost price is enough to guarantee a reasonable return for storage and other costs of owning the grain. A spread equal to 3 cents a bushel a month is usually considered profitable. The profitability of a spread will fluctuate with price levels and interest rates. For example, if in September the December futures price is selling at 9 cents over the terminal spot cash price of that same grade, it would probably be considered worthwhile to buy this grade of grain at country points and to store it with the intention of selling a like quantity on futures market. Assuming in this hypothetical situation that there are no sudden price moves to throw the spread out of line, in December the merchandiser would be able to use his stored grain to deliver on his futures obligation at the full 9-cent profit.

As is true of other hedgers, the country elevator operator is not protected against unfavorable movements of the basis, and he must use judgment and care in his hedging operations. To hedge most efficiently the country elevator operator should carefully select the futures contract that bears the closest relationship to the grain to be hedged. He usually sells futures whose delivery month is closest to the time at which he expects to sell the grain because the price for that month will probably bear the closest relationship to the cash price he will receive for the grain he sells. This is not a hard-and-fast rule because conditions do change. For instance, at times it may be desirable for an elevator operator to transfer hedges from one month to another, although it is a matter of keen judgment to determine just when this transfer, or *switch,* should be made.

Financing Grain Purchases

Because the purchasing and handling operations of the country elevator operator involve cash payment to the farmer for his grain upon delivery, it would appear that, in periods of high grain prices and heavy movements, tremendous capital would be required. Actually, however, the required amount of capital is small when compared to the volume of business handled. For example, an elevator operator may buy 1,200 bushels of grain from a farmer at $2.50 a bushel and pay the farmer $3,000 upon delivery. Having previously ordered cars from the railroad agent, the operator has the grain loaded while it is being delivered and, shortly after the last truckload arrives, he knows within a few bushels how much the railroad car contains. If the grain is already sold, he also knows what his approximate return on the car will be. The next step is to fill out the shipper's bill of lading on forms supplied by the railroad, stating weight, destination, and other pertinent items.

After the railroad freight agent has accepted the car, it is sealed. The seal bears a number for identification and legal protection. The agent for the railroad accepts the bill of lading by signing it and releases the car for transportation. The elevator operator may then take what is known as a *bill of lading* to his local bank, endorse it to make it negotiable, and draw a demand draft on the consignee for the major portion of the money invested in the grain. After the draft is drawn and the bill of lading endorsed, giving the bank control of the grain, the bank will credit the elevator operator's account with the amount of the draft. The bank then sends the bill of lading with the draft attached to a bank in the consignee's city. After the draft is presented to the consignee and paid by him, the bill of lading is released to him. If, however, the shipper does not need to borrow money on the shipment, he requests a straight bill of lading, which is made out directly to the consignee. The straight bill of lading is mailed by the shipper with an invoice to the buyer or commission man at the point of destination, and the shipper awaits payment by draft, check, or credit to his account at the time of final unloading and settlement.

This credit system is of benefit to the average shipper chiefly in supplying him with capital and occasionally in serving to protect him against the necessity of extending credit to the consignee. For instance, if the shipper draws a draft against a small buyer who does not have the credit rating of the established terminal market buyer, he is protected because the draft goes with the bill of lading to the consignee's bank, and the consignee must arrange to pay the draft before he can obtain the bill of lading entitling him to the grain.

It is a relatively simple matter to finance the country grain business when the grain is sold promptly after purchase. If, however, the operator decides to store it, he can hedge the purchase by the sale of futures contracts, thus eliminating the price risk to the extent that a commercial bank will usually undertake the necessary loan.

Other Sources of Income

In addition to the income for storing and handling, an elevator operator can sometimes increase his profit by conditioning grain in order to improve its grade. Conditioning operations include cleaning, separating, mixing, and drying. Elevators equipped with cleaners and separators have many opportunities to improve the grade by removing unwanted material from grain.

As an example of mixing, assume corn receipts at an elevator are grading No. 3 because of their moisture content but are entirely free from damaged kernels, while at the same time a specific delivery of corn qualifies as No. 3 on moisture but because of damaged kernels is lowered to No. 4 grade. By blending a part of this latter with the corn grading No. 3 on account of moisture but containing practically no damaged kernels, the amount of damaged kernels in the mixed lot would be reduced enough so the entire lot of corn would grade No. 3. By this operation the value of the car of corn might be increased, making it possible for the country elevator to pay a better price to the producer and to receive a better price from his buyer.

Grain at country elevators often has more than one grade-determining factor, such as moisture or damaged kernels; removal of a small percentage of one of these factors, such as moisture, does not always improve the grade because of the predominance of the other grade-determining factors. An increasing number of country elevators are now equipped with dryers, however, a large proportion of the grain received at country elevators is forwarded to terminal markets in much the same condition in which it is received because normally the run of grain from a given locality is rather uniform in grade characteristics. The grain may also be of a type that cannot be improved profitably by mixing, cleaning, or drying.

Another source of income for the country elevator is the merchandising of materials required by farmer customers. Today, many elevators operate a well-stocked lumberyard that handles such products as building materials, fencing, paints, building hardware, posts, shingles, and fuel.

Another important adjunct of many modern elevators is a feed department. In most cases, elevators handle commercially mixed feeds and, in some cases, such feed ingredients as cottonseed meal, linseed meal, soybean meal, bran, and middlings. In areas where livestock are fed, the elevator will usually have a feed mill where farmers' grains can be ground and mixed with other feed ingredients. Under the direction of a competent feed man, an elevator may also produce its own line of feeds according to its own formulas and sacked under its own trade name.

Many elevators sell seeds, some handling seeds of a commercial seed company, others buying local seeds and cleaning them for resale. Other items frequently handled by country elevators include fertilizers, fumigants, and twine. Several factors have contributed to this trend toward increased diversification, including the smaller profit margin in grain handling and the decrease in the number of smaller elevators due to the development of larger and more conveniently located units.

River Elevators and Subterminals

While this chapter deals primarily with country elevators, the role of the river elevator and the subterminal should be mentioned. Originally these elevators were operated almost exclusively as feeders for terminal elevators, although in recent years, due to the growth of the poultry industry in the South, as well as the demand for barge grain at the Gulf for export, most of the grain originating on the water does not go to the terminal house.

Many river elevators have no rail facilities and rely on grain brought by truck from country elevators in the vicinity. Grain is run through the house in the usual manner but is loaded on barges rather than boxcars for towing to the terminal house, export elevator, or other consumer. While some barges are owned by the elevator or its subsidiaries, most barges are hired from barge companies.

A subterminal elevator is typically located apart from a metropolitan area and is often the only large grain-handling facility in its immediate vicinity that does not buy grain directly from the farmer. Bushel capacity varies greatly among subterminal elevators; some subterminals are smaller than the largest country elevator, while others are larger than some terminal houses.

The manager of a subterminal elevator merchandises directly to terminal elevators, processors, and exporters rather than selling to interior dealers or commission merchants. The facilities and operations of a subterminal are similar in many ways to those of the terminal house.

8

cash grain merchandisers

The term *cash merchandiser* or *cash merchant* is used to describe any person or firm dealing in the buying and selling of a cash commodity. A cash grain merchandiser operates in the cash market as both a buyer and seller of grain. He acts as his own principal and his income is derived from changes in the basis.

The cash merchandiser may operate in a deficit grain-producing area, buying from a surplus area to supply the deficit area; or he may operate in a surplus grain-producing area, buying grain from this area and selling it in the deficit areas. The cash merchandiser in surplus areas operates between the country elevator and the terminal elevator, exporter, processor, miller, and other buyers of grain.

Types of Cash Grain Merchandisers

Cash grain merchandisers can be classified into three types. One is the *country merchandiser,* who is generally located away from a terminal market. The country merchandiser buys grain from country elevators and sells it to terminal markets, to grain processors, and to exporters. He does not operate storage facilities, although he does take title to the grain he buys. The second type is the *terminal market merchandiser,* who buys grain (usually from country elevators) to be received in a terminal market, such as Chicago or Minneapolis. The terminal market merchandiser, like the country merchandiser, does not operate or own storage space, although he also takes title to the grain he buys. The third type is the *terminal elevator company,* which participates in the same merchandising operations as other cash grain merchandisers in addition to operating storage facilities at terminals and subterminals. The merchandising techniques of terminal elevators will be discussed in a later chapter.

Commission Brokers

Although commission brokers are sometimes considered cash grain merchants, they differ from the three basic types of grain merchandisers because they do not take legal title to grain. They function strictly as agents and are usually paid commission fees by the seller of grain. After a broker has brought the buyer and seller together and the terms and conditions of the sale are mutually agreed upon, the brokerage service has been completed. The major role of the commission broker in grain marketing today is in the movement of fats and oils (lard, soybean oil, cottonseed oil, etc.), millfeeds, protein meals, and by-products of various processing operations. The commission broker also handles volume movements of grains and soybeans such as 100 hopper-car trains, barges, and vessels.

Operations of the Cash Merchandiser for Country Elevators

The cash merchandiser is the major link between the country elevator and the later steps in the marketing process. Since a country elevator must pay the highest price possible to compete successfully with other elevators for farmers' grain, the most important service performed by the merchandiser is finding the highest bidder for country grain. Country elevator operators usually use the services of a cash merchandiser to shop and compare the bids of the various users and/or markets for grain, although some country shippers do sell directly to processors or terminal elevators.

The cash merchandiser is a specialist in serving the needs of country elevator management. He contacts many country elevators (the number depending on the size of his operation). Through his contacts, the cash merchandiser is able to provide his shippers with information on both cash and futures markets, and he may handle their transactions in the futures market.

The merchandiser may help the shipper in obtaining empty railroad cars or may even operate his own fleet of hopper cars. Because of almost daily personal contacts with railroad personnel, the merchandiser may also be able to expedite the placement of empties and trace the movement of railroads cars in transit to expedite the haul. When grain is lost due to railroad error or grain quality is diminished due to delay in transit, the merchandiser can prepare or obtain the necessary claims on behalf of the shipper. On this evidence, the railroad will make or refuse a cash settlement with the shipper.

At terminal markets, the merchandiser is given a sample of the grain from each railroad car billed or shipped to him. On the basis of his knowledge, he may advise the shipper to call for an appeal inspection if he believes the car has been misgraded. In such a case, the merchandiser either applies the carload to the best applicable contract or sells it on the open market for the shipper at the best price available.

The financing of country grain is secured by the country elevator's bill of lading. Shipper's order bills of lading with a draft attached are presented to the merchandiser for payment by a bank in the merchandiser's city. When the merchandiser pays the draft, the bank releases the bill of lading to him, and legal title passes. In some cases, the country shipper will mail the merchandiser a bill of lading (either order or straight) and request a cash advance amounting to approximately 80% of the value of the car. Regardless of the method used, the capital requirements of the merchandiser can be large. At times, he will borrow money to pay drafts and requests for advances from country shippers and then immediately draft through banks or request advances from his buyer to repay his bank loan or recoup his capital.

Operations of a Terminal-market Merchandiser

A normal day for a terminal-market merchandiser begins after the close of the market when he records the overnight bids of the users of grain. These bids are the foundation for his own bids to country shippers. He also notes representative billing and spot sales for various types and quality to insure that he has full and complete knowledge of the day's developments. This information is contained in the daily records maintained by the various grain exchanges. Disposition for the cars handled during the day must be obtained from his buyers so that orders can be given to the railroads for movement to destination.

Bids to country shippers can be on a to-arrive basis or on an on-track basis, depending mostly on the preference of the buyers. To illustrate: assume the best bid for corn today is $2.50, delivered Chicago by rail, and the corn rail rate from E. Overshoe, Illinois, to Chicago is 15 cents a bushel. The merchandiser could pay the elevator operator at E. Overshoe $2.50 delivered Chicago or $2.35 on track (f.o.b. E. Overshoe). If the country shipper decides to sell grain on track at $2.35 the merchandiser would collect the 15 cents freight from his buyer since it is included as part of the $2.50 price. If the shipper sold delivered at $2.50 he collects the 15 cents freight in the price. Thus the E. Overshoe elevator operator will net the same price for his grain regardless of whether he sells delivered or on track, provided there is no change in quality from the time the car was loaded until the delivery at destination.

In recent years, a much greater portion of country grain has moved by truck. Sharp increases in rail grain rates, chronic shortages of railroad cars, speedier harvests that quickly congest country elevator facilities, and a pronounced demand for truck grain by some users have all encouraged truck movements of grain. In response, the terminal market merchandiser has provided country shippers with bids for grain to be delivered by truck. In general, the mechanics of trading truck grain closely follow those described for rail grain.

Basis Trading by the Merchandiser

A merchandiser must evaluate the current basis in terms of typical basis patterns and the structure of the futures market. Because profits from basis trading are the biggest contributor to the total income of the merchandising firm, basis patterns are closely studied. The basis generally shows a certain seasonal pattern, i.e., it is at its widest discount or smallest premium to the futures about a month after the harvest and generally shows improvement over the next eight or nine months.

As an example, assume a carrying charge market for corn on December 1:

December	March	May	July
$2.18	$2.22	$2.26	$2.30

If cash corn on December 1 is bid at the December futures price, then at the same point in time the basis could also be quoted as 4 cents under March or 8 cents under May or 12 cents under July. The typical tendency of the basis has been to adjust to the general basis level of the preceding expired future. If the basis in December is December price, and the December future expires, then, probably before the March future expires, the cash basis will trade near March price. With the March futures expiration, the basis will tend to work upward to May price. The same situation will tend to occur when the May future expires and cash is based on the July.

Basis trading is strongly influenced by the structure of the futures market—whether carrying charge or inverted. Being *long the basis* (bought cash, sold futures, expecting the basis to improve) is less risky in a carrying-charge market than in an inverted futures market. Being *short the basis* (sold cash, bought futures, expecting the basis to worsen) is more risky in a carrying charge market than in an inverted futures market. Being long the basis in an inverted market has similar risks to being speculatively long an inverted futures delivery month. Being short the basis in an inverted market is less risky than being speculatively short in an inverted delivery futures month because the inverse itself tends to stabilize or cause the basis to worsen.

Bull markets in futures are usually accompanied by a tendency to carrying charges and a cash basis that follows usual patterns. The upward thrust of futures prices is incentive enough to attract cash grain. Bear markets are also usually accompanied by carrying charges. A declining futures market, however, may tend to tighten or improve the cash basis because of the price resistance or lack of selling of cash grain at low levels.

While merchandisers may be either long or short the basis, some are consistently long the basis (short hedgers) and others are consistently short the basis (long hedgers). The option of being either long or short the basis allows the merchandiser great flexibility in his operations.

When the merchandiser is working on a basis trade, he may be able to offer the country shipper bids equal to or higher than the bids from processors, millers, terminal elevators, and other buyers. As an example, assume a merchandiser was short the basis at 4 cents over last week, and today the best basis bid is only 1 cent over. To cover the 4 cents over sale, the merchandiser might be willing to pay 1 cent or even 1½ cents over, and while this might appear to be a strong basis, it would still leave a basis profit of 2½ cents or 3 cents per bushel.

Records Maintained by the Merchandiser

To be effective the merchandiser must keep and maintain accurate records of purchases and sales. One important part of his purchase record is the record of freight billing. The accurate maintenance of this is vital in providing accurate knowledge of his purchases compared to his obligations to provide certain specified billing to cover his sales. The merchandiser in the terminal market is responsible for the payment of all inbound freight cars and the ordering of freight cars to their outbound destinations from the inspection tracks. He must also keep accurate records on all freight due and cars ordered.

Another important record is the long-and-short-position record. This has three sections—*cash, futures,* and *net.* All purchases and sales of cash and futures are recorded, the difference representing the net or unhedged portion. The merchandiser is usually fully hedged, even though he may have a sizable cash position, long or short, and a large offsetting futures position, either long or short. The merchandiser normally attempts to stay net even, content to make a profit on basis trading rather than being speculatively net long or short.

9

terminal elevators - operations

Terminal elevators perform a variety of services to the grain industry, including storage of grain between the time of harvest and/or shipment from country points until it is needed; conservation or improvement of grain quality; merchandising to exporters, processors, or other domestic users; and facilitation of the orderly functioning of futures markets.

Terminal elevators, as the name implies, are located in the principal grain-marketing centers and may vary in capacity from a few hundred thousand bushels to over 30 million bushels. The location of the elevator rather than its size determines its classification. The terminal elevator operator buys grain from cash grain merchants, subterminal elevators, river houses, and country elevators. Depending on the facilities of the terminal elevator, grain may be received by rail, truck, barge, or, sometimes, by boat. The terminal elevator, in turn, sells grain to processors, millers, distillers, feed manufacturers, exporters, and, sometimes, to elevators in other parts of the country. The elevator may provide for shipment by rail, water vessel, or barge, or, on rare occasions, may sell small quantities for truck shipment.

Types of Terminal Elevators

The classification of terminal elevators according to ownership is much like that of country elevators: (1) *independent,* (2) *cooperative,* and (3) *integrated.* The independent and cooperative companies are similar in organization to the country elevators so designated. The integrated terminal company operates terminal elevators in more than one marketing center. The company may own line country elevators, river houses, and subterminals; it may be engaged in exporting; it may lease rail equipment; and it may own barges and boats. While overall policy, as well as financing of the integrated terminal company, may be provided by the home office, day-to-day merchandising activities are carried on by the individual management at each terminal location.

Physical Layout

Physically, all elevators may be divided into two parts; the workhouse and the storage annex or tanks. The workhouse contains the machinery for unloading, loading, weighing, conditioning, and moving grain within the elevator. The storage annex is the grain storage area and is connected by conveyor or belts with the workhouse. The belts move grain horizontally and are located either beneath the tanks or bins (shipping belts) or above the tanks (receiving belts). Over these belts, grain is received, moved within the elevator, and shipped out.

Receiving and Handling

One way to illustrate the operation of an elevator is to follow the movement of grain through it. If the grain arrives by boxcar, it is sampled on the elevator's tracks by elevator personnel to insure the quality is equal to the official grade on the car. If the official grade checks out, the car will be *spotted* (positioned) for unloading; otherwise, an appeal inspection may be called and the car pushed by the unloading dump, pending the resampling.

At modern plants, grain is unloaded by a *car dumper,* which is a movable rail platform. After the car has been spotted, it is clamped to a car dumper next to the unloading pit or sink. The railcar is then mechanically tilted up or down and to the side, allowing the grain to flow out of the car into the unloading pit. Normally, a car can be unloaded by this method in from six to eight minutes, at the rate of approximately ten cars per hour. At large facilities, more than one car dumper may be operating.

Another unloading method is the power shovel. It is used by smaller elevators that do not have sufficient volume to justify the high initial cost of a car dumper. At a power-shovel plant, the boxcar is spotted over the unloading pit, and the grain doors are removed to allow the free flow of grain out of the car. Power shovels are 3 feet by 4 feet, have two handles, and are made of either heavy board or aluminum. A cable powered by an electric motor is attached to the bottom front side. Using a clutch device on the cable, the power-shovel operator moves back to the end of the car and engages the clutch, and the motor pulls the shovel to the door of the car, discharging the grain into the pit. This procedure is repeated until the car is emptied. Usually two shovelers can unload a car in about 30 minutes. Cars that are too long for a car dumper must be unloaded by power shovel.

If the grain arrives by hopper car, no unloading equipment is required. Hopper cars are spotted over the unloading pit, the doors or traps are opened, and the grain runs out the bottom of the car. Once unloaded, cars are moved away from the unloading pit by winch cable attached to an electric motor, by motorized power units that ride the rails and shuttle the empties off the dump, or by diesel locomotives.

When the grain comes to an elevator by truck, the truck is driven onto the unloading platform and secured. After the tailgates have been removed or opened, the front end of the truck is raised, and the grain flows into the unloading sink by gravity.

The grain dumped into the unloading pit is carried by a belt to the leg boot, the lowest part of the *leg* (an enclosed, vertical endless belt which moves grain vertically in the elevator). V-shaped metal buckets riveted to the leg scoop up the grain, carry it from the boot to the *head* (the topmost vertical point), and spill it into a spout that loads to the receiving *garner* (a bin that collects the entire contents of a car or truck before it is dumped into the hopper scale for weighing). After being weighed, grain is dropped by a spout onto a receiving belt and moved to a point where a movable device called a *tripper* diverts the grain into the desired bin for storage.

Unloading grain from boats or barges is accomplished by use of a marine leg, which is movable so it can be positioned or dipped into the grain in the vessel's hold. The buckets of the marine leg scoop up grain and discharge it by spout to a belt that conveys the grain to a fixed leg in the elevator. When grain will no longer run to the marine leg, it is pushed to the leg by automatic or manually operated power shovels, and then the hold is swept. From this point, the grain is handled in the same way as previously described for boxcars and trucks.

Management of Stocks

The superintendent of the elevator maintains a very complete and up-to-date record of all grain in his elevator. Much of the information is contained on a blackboard diagram locked in the elevator's office. Grain coming into the elevator is *binned* (put into a storage tank) with grain of similar characteristics; this decision is usually made from data shown on the superintendent's board. The data includes the class, subclass, and grades, showing all the grading factors of the grain in each bin, the number of bushels of grain in each bin, the date the grain was put into storage, and the origin of the grain.

Temperature-recording systems are an important aid to the superintendent in avoiding deterioration of grain in storage. These systems, installed in each tank or bin, register the temperature of grain at various levels by means of a cable. Temperature readings are taken and recorded regularly, so that any sharp temperature increase alerts the superintendent to the possibility that a bin or part of a bin is in danger of going out of condition.

Conditioning and Quality Improvement

Operations to condition grain include running, drying, and cleaning. *Running* or *turning,* which is done to cool and aerate grain in danger of deterioration, is accomplished by moving the grain over belts or through spouts. *Drying* is the removal of enough excess moisture to permit safe storage. The variety of mechanical driers available superheat air by means of either natural gas, fuel oil, or steam coils. The heated air is forced up through the mass of grain, removing moisture as it rises. The grain is then cooled and binned. *Cleaning* or *aspirating* removes unwanted material from grain by the use of air blowers, screens, and cylinders. Cleaning may also be used to improve the quality of grain.

Other operations to improve the quality of grain include mixing or blending, clipping, plumping, washing, and scouring. *Mixing* or *blending* is the combining of two or more lots of grain in an effort to make a new lot that will be of better quality than at least one of the lots entering the blend. The superintendent's board provides the necessary information on which lots will be combined. *Clipping* is performed to increase the test weight of oats or barley. A machine called a *clipper* is used to knock off the ends of the hull, thereby raising the test weight of the kernel. *Plumping* is accomplished by the use of a needle machine, which removes the *needles* (thin kernels) from oats and rye. This increases the test weight and makes the remaining berries much more uniform in size. *Washing* and *scouring* are performed only rarely and then only to remove smut from wheat and oats. Specialized machinery is required for each operation.

Loading Grain

Equipment used in loading grain out of the elevator is similar to that previously described. The first step is to tap the desired bin by opening the movable slide at the bottom of the bin or tank, thereby allowing the grain to spout or flow onto the shipping belt. The shipping belt carries the grain to the boot of the leg, which then elevates the grain to the head, where it is spilled by spout to the shipping garner and later to a hopper scale. When the proper quantity is weighed in the scale, the grain is dropped through a spout directly into the boxcar with the grain doors in place, or into the truck.

When the grain is to be loaded into boats or barges, however, it is dropped from the hopper scale into shipping bins (located on the waterside of the elevator) rather than spouted directly into the vessel's hold. By closing the spout slides of the shipping bins, loading can be stopped to allow the grain on the boat to be *trimmed* (leveled) to make it seaworthy. While the load is being trimmed and the grain is being held in these bins, all operations in the elevator itself may be continued.

Supervision of Elevators

The extent of third-party supervision of elevator operations depends on the degree of public involvement in them. There are two general classes of elevators, private and public or semipublic. *Private elevators* are usually owned by processors or millers who use them for storing and conditioning grain used only in their own processes. Except to insure that grain is properly weighed into an elevator, no other public involvement exists for private elevators.

Public or semipublic elevators are public utilities, and as such their facilities are available to the public at published tariff rates. To accommodate this public interest, these elevators are federally licensed under the provisions of the U.S. Warehouse Act. If all of the facility is available for public use, it is called a public elevator; if only part of the facility is for public use, it is called semipublic. Because public demand for elevator facilities is limited, as a practical matter, public or semipublic elevators are merchandising elevators with most of the grain owned by the operator.

Many public elevators are regular for delivery on futures contracts. To become regular a house must not only be federally licensed, but it must also meet commodity exchange standards safeguarding the taker of grain on delivery. The conditions that determine regularity of elevators, warehouses, and shipping facilities are part of the rules of the commodity exchange. In order to protect the holder of the warehouse receipt or shipping certificate, such facilities must post bond with the exchange after approval of regularity is given.

In Chicago, weighing is performed by grain-elevator personnel who are licensed and bonded under the provisions of the U.S. Warehouse Act. The Chicago Board of Trade Weighmaster has the additional title of Board of Trade Registrar and is in charge of registering and cancelling warehouse receipts for houses regular for delivery on futures contracts.

A. Weighing Supervision

Years ago, grain was bought and sold in units of volume measured by the bushel or its subdivisions. But this basis of accounting for quantity was subject to certain inaccuracies. Exact measurement was difficult to obtain because of the physical nature of the grain. Furthermore, as the grain trade increased in volume and complexity, buyers and sellers could no longer both be present to determine the accuracy of the quantity under consideration in a particular transaction. Hence, the practice of weighing grain instead of measuring it was inaugurated early in the development of the Chicago Board of Trade.

Grain weight is based on the volume measured by the Winchester bushel, so it varies somewhat with the different grains. In grain-market parlance, today a bushel is understood to mean 60 pounds for wheat and soybeans, 56 pounds for corn and rye, 48 pounds for barley, and 32 pounds for oats. In other words, in considering a specific quantity of grain, the bushel is an arbitrary unit of weight rather than of volume.

The difficulties of accounting for quantity, however, were not solved by the exchange from volume measurement to weight. Scales varied in accuracy. Pilferage and leakage might occur in transit and endless disputes might result. With the buyer and seller miles apart, some independent and impartial means for determining the accuracy of the weighing was needed.

The need for impartial weighing of grain was apparent to those engaged in organizing the Chicago Board of Trade. The Act passed by the Legislature of Illinois in 1859 to incorporate the Board of Trade included the following provision:

"Said corporation shall have power to appoint one or more persons, as they may see fit, to examine, measure, weigh, gauge, or inspect flour, grain, provisions, liquor, lumber, or any other articles of produce or traffic commonly dealt in by the members of said corporation; and the certificate of such person or inspector as to the quality or quantity of any such article, or their brand or mark upon it, or upon any package containing such article, shall be evidence between buyer and seller of the quantity, grade, or quality of the same, and shall be binding upon the member of said corporation, or others interested, and requiring or assenting to the employment of such weighers, measurers, gaugers, or inspectors; nothing herein contained, however, shall compel the employment, by anyone, of any such appointee."

Although the Weighing and Registration Department of the Chicago Board of Trade was organized to carry on this impartial weighing in the Chicago district, the work of the department was handicapped at first by lack of uniform practice between Chicago and some other markets and areas, which caused frequent misunderstandings. As a result, a conference of weighmasters of the principal terminal grain markets in 1918 adopted a resolution defining the following standard weighing practices:

1. Supervision sufficient to assure intelligent inspection of all boxcars, trucks, and boats/barges inbound and outbound, and inspection of the delivery of all grain from carrier to scale or from scale to carrier

2. Inspection of scales and equipment used to weigh and handle grain to and from grain carriers

3. Comprehensive records of all activities

4. Sufficient authority on the part of the weighing staff to enforce rules and regulations recognized to be essential in order to protect properly the weights of grain

The adoption of this resolution did much to establish not only uniformity but also high standards of weighing and scale-inspection performance throughout the grain trade.

B. The Weighing Procedure

When a boxcar, truck, or boat is spotted for unloading, the deputy weighman has a number of duties to perform before ascertaining the precise weight of the grain it carries. First, he records the results of an examination of the carrier, including the numbers of any protecting seals, and notes any leakages. He then examines and balances the scale. After assuring himself that all elevator equipment is in good order and grain-tight, he sees that the weighing is done without delay. Extra handling of grain before weighing or the use of attachments for the purpose of blowing and cleaning grain that might in any way affect its weight are not permitted, nor is any other condition tolerated that might be conducive to shortage. It is the duty of the deputy weighman to prevent waste and to see that the grain carrier is thoroughly swept after unloading.

On outbound shipments, licensed and bonded elevator personnel first determine whether the conveyance is suitable for grain loading. Secondly, after an examination of elevator equipment and scales, the weighman insures that prompt loading takes place. After loading, protecting seals are placed, when applicable.

To guard against error, the weights are verified with each use by a recording device, or by recording the denominations of the hanger-weights used on the scales. Original records are entered in permanently bound books. Finally, the deputy weighman sends a complete report covering each step in the process of handling every lot. This comprises a case history and is placed on file in the Weighmaster's office. An official weight certificate is provided the owner of each lot of grain received at or shipped from the terminal market.

The certificates of weight and/or the weight records issued by the Weighmaster have the following uses:

1. As a basis for determining the quantity of grain in settlements between the shipper and the buyer

2. To determine the amount of grain being delivered and received on a futures contract

3. As the basis of payment to the inbound carrier for freight charges

4. In substantiation of claims against a carrier for weight losses occurring in transit, either to or from the market

5. As the basis of payment to the outbound carrier

C. Supervision of Federal Elevators Not Regular for Delivery

Not all states have laws governing warehouse operations, and, in some states with warehouse legislation, effective regulation does not always prevail. Regulations provided in the U.S. Warehouse Act, however, require a federal warehouse to be bonded, to account for all warehouse receipts issued to it by the U.S. Warehouse Branch of the USDA, and to submit to federal examination of books, records, and actual inventories. Because of these safeguards, negotiable federal warehouse receipts enjoy high prestige as collateral in financial quarters.

D. Supervision of Federal Elevators Regular for Delivery

Previous comments on federal houses not regular for delivery apply also to federal houses regular for delivery, although regular elevators encounter still more regulation.

After grain is received at an elevator, a federal warehouse receipt is prepared by the elevator company showing the numerical grade of the grain and the weight. To substantiate this information, a copy of the inspection certificate and of the bill of lading receipt accompanies the receipt when it is taken to the Exchange Registrar for registration. The receipt can be delivered against a futures contract only after registration.

Before grain can be loaded out of a regular elevator, sufficient warehouse receipts covering the weight and grades to be loaded must be cancelled by the Registrar. Even grain movements within the elevator that alter weights or grades may not take place without cancellation of appropriate receipts. After the cancellation has been recorded by the Board of Trade Weighmaster, the deputy weighman at the appropriate elevator is notified to permit the cancelled quantity to be loaded out.

E. U.S. Grain Standards Act of 1976

Late in 1976, PL 94-582, known as the U.S. Grain Standards Act of 1976, was passed by Congress. The effective date of the new legislation was November 20, 1976, 30 days following its enactment. The new legislation set up the Federal Grain Inspection Service which is empowered to establish standards of weighing and weight certification, grading, and transportation of grain in interstate and foreign commerce. The Service is empowered to establish standards of kind, class, quality, and condition of grains.

10

terminal elevators - merchandising

Most of the grain bought by a terminal elevator is bought without a specific buyer in mind, but purchases are made only after consideration of the basis; the billing; the possibility of profit from storage, blending, and/or conditioning; and the availability of storage and the level of demand.

Analyzing Basis

Terminal elevator merchandisers tend to think in terms of buying and selling basis differences, the difference between the cash and futures prices, rather than in terms of buying and selling at flat prices per bushel. Several factors enter into an analysis of basis. The first of these is the usual seasonal relationship between cash and futures prices. The cash price tends to be lowest relative to the futures price during the middle and late stages of the harvest. This is a period of *weak* basis and reflects usually ample, perhaps burdensome, supplies and, at times, a tightness of storage space. Earlier in the harvest, processors and millers may be aggressive buyers of grain because of scarce stocks at the end of the old-crop year. After this initial demand has been satisfied by newly harvested stocks, however, the basis will usually weaken as cash prices will decline relative to futures. Often, the weakest basis, lowest cash prices relative to futures prices, occurs from a month to six weeks after the harvesting peak when storage space is fairly well filled. From this point on, the basis usually tends to improve, that is, cash prices rise relative to the futures.

Terminal elevator merchandisers usually begin buying grain for accumulation at or near the peak marketing period coinciding with harvesting and will normally accelerate their buying as the basis weakens, that is, as the cash price declines relative to the futures. Employing this kind of strategy will normally result in the terminal elevator's weighted-average buying price being near the low for the crop harvesting period. However, basis analysis must also take into consideration the expected movements of the futures prices in later periods. In periods when the outlook is for a protracted bear market the basis is likely to be wide, the cash price well under the futures price. Conversely, the outlook for an extended bull market tends to cause a steady or more narrow basis, with cash prices closer to the futures. The outlook for carrying charges and possible changes in carrying charges must also be analyzed in the framework of overall supply and demand by the terminal elevator merchandiser in accumulating stocks of grain.

The second factor in basis analysis is the amount of empty storage available in elevators. Several fixed-cost elements, such as taxes and depreciation, exist for all elevators whether there is grain in the elevator or not. Even though grain may not be purchased at a basis as favorable as desired, having grain in the elevator may make it possible for the elevator to earn the variable costs, possibly part of the fixed costs, and even a profit. Empty storage capacity, then, may encourage merchandisers to buy grain at a basis that is not necessarily attractive.

The third factor in analyzing basis is the merchandising opportunities that may exist for current grain stocks. As an example, assume that a stock of wheat accumulated at an average basis of 2½ cents under the current futures price can be sold at 3½ over the futures price. Assume further that the wheat merchandiser feels that this basis is not likely to improve significantly over the next two months. In such a case, if corn could be purchased at a basis of 3 cents under the futures, the merchandiser might well decide to sell the wheat earning 6 cents on the basis and fill the elevator with corn.

Final factors in analyzing basis are the existing cash price and futures price considered in light of the grade differentials and corresponding premiums or discounts for delivery on futures contracts. For those elevators *regular,* approved, for delivery on futures contracts, with grain of deliverable grade, in a period in which carrying charges exist, delivery against futures contracts sold would be viable. Not all these conditions exist at any given time, however. There are occasions when delivery of grain stocks against short futures positions can provide elevators with significant revenue.

As examples of an elevator's considerations in deciding whether or not to deliver stocks against short futures contracts, examine the following situations:

Situation 1
A terminal elevator approved for delivery owns No. 2 Yellow corn bought at an average basis of 5 cents over the current futures, and cash corn can be sold at 10 cents over the current futures.

Situation 2
A terminal elevator approved for delivery owns cash corn bought at 4½ cents over the current futures, but cash corn is only worth the current or nearby futures price.

Situation 3
A terminal elevator approved for delivery owns cash corn bought at an average of 5 cents under the current futures and corn can be sold on the cash market at 1 cent over the current futures.

Under these situations, would delivery against a short futures position prove profitable?

Situation 1
The elevator has a 5 cent basis or trading profit. No. 2 Yellow corn is deliverable at contract price so the elevator would lose 5 cents by delivering against the futures.

Situation 2
This elevator operator has a position reflecting a net basis or trading loss of 4½ cents. How long the elevator operator is willing to live with a losing position will largely determine whether delivery against the futures will be made. If the elevator operator chooses to liquidate the position by delivery against the short futures position, it can be done with a small gross profit, since No. 2 Yellow corn can be delivered at contract price plus storage charges until load-out and elevation (loading out) charges. This of course would make very little contribution to costs, but delivering does represent a more palatable alternative than the prospect of 4½ cent basis or net trading loss.

Situation 3
Delivering against the futures would represent the best possibility, since the cash corn could be delivered equal to 5 cents over (contract price plus, say, 5 cents elevation charge), which would provide a 10-cent profit compared to a 6-cent basis or net trading profit on corn sold on the cash market.

Billing

Terminal merchandising requires a working knowledge of transportation fundamentals. Transit privileges are available to most terminal elevator operators. Grain received by trucks is flat or nontransit tonnage, i.e., it cannot be applied against an outbound freight bill without a penalty. Grain received by barge or rail is transit tonnage, i.e., it can be applied against an outbound freight bill. At no time can inbound barge or rail billing registered for transit purposes exceed the amount of grain in the elevator. Outbound grain shipments by boat, however, are not affected by any inbound transit considerations.

The most expeditious use of truck or flat billing is against a boat sale. Occasionally flat billing will be applied against a rail sale, but a flat (or point-to-point) rate (usually higher) must be applied from the terminal elevator to destination instead of a proportional or reshipping rate, which is usually lower. Since terminal elevators usually sell on a delivered basis, the terminal operator may be obliged to quote a price to compensate for the higher flat rate, thereby making the delivered price higher than competitive offerings based on transit tonnage. Inbound barge billing may be applied against an outbound rail movement but should not be applied against an outbound vessel movement, because the amount paid in billing may be lost.

A situation of inbound transit tonnage being shipped outbound by a nontransit carrier gives rise to a transportation technique called *tonnage* or *billing cancellation.* This is accomplished by removing from a transit record certain inbound billing originally recorded in anticipation of outbound transit application. Billing cancellation can, of course, cost an elevator operator money, because funds have been expended for the original inbound bill. Consequently, billing-cancellation losses must be considered when boat sales are made if adequate flat billing is not available.

With the development of the St. Lawrence Seaway and the resulting large boat movements of grain, some terminal operators in port locations cannot buy enough truck or flat billing to equal boat sales, so billing cancellation must be calculated as a cost when offering grain for boat sales. Another solution to this problem is for such terminals to change their buying methods, both by mode of transportation and by price differentiation, to more closely correlate with the expected outbound modes.

Chicago elevators choosing to make delivery against a short futures position must apply inbound transit billing against the outbound shipment for the warehouse receipt owner sufficient to protect the minimum inbound proportional rate to Chicago. Because of the deficit of flat billing and the need to pay flat billing to protect the proportional rate, seldom, if ever, would truck billing be applied against a warehouse receipt shipment, because the penalty would be prohibitive.

Blending and Conditioning

The third important consideration for elevator operators is the possibility of profits or losses from blending or conditioning. The merchandiser must have an intimate knowledge of the grades of grain in his elevator. When spot grain is offered to him he must be able to analyze the prospects for a blending profit. *Blending* is a mechanical procedure of combining two or more lots of grain having different quality factors in an effort to produce a lot of grain that will be of better quality than at least one of the lots entering the blend.

As an example, assume two bins each containing equal quantities of corn. One bin contains No. 2 Yellow corn, with the following factors: 55 pounds test weight, 14% moisture, and 3% damaged kernels. The other bin contains No. 4 Yellow corn with the following factors: 55 pounds test weight, 18% moisture, and 1% damaged kernels. If equal quantities from each bin are run together the following factors should emerge: 55 pounds test weight, 16% moisture, 2% damaged kernels; this blend would be graded No. 3 Yellow corn. The No. 2 Yellow corn was bought with no discounts, but the No. 4 Yellow corn was bought with discounts of 5 cents per bushel because the moisture was in excess of 15½%. Average discounts applicable on these two lots of grain would be 2½ cents per bushel (0 + 5 = 5; 5 ÷ 2 = 2½ cents). Since the resulting lot of grain would carry a discount of 1 cent because of excess moisture, however, only 1½ cents gross blending profit would result from the 2½ cents original average discount.

The preceding example is a simple illustration of blending. In the larger elevators, blends of three or four different lots are not uncommon, and blending against all factors in the grade may take place. Blending is a legitimate function, not an adulteration, since grain is usually sold by grade. Grades are determined by a disinterested, impartial third party, and the grade certificate fully states all pertinent quality factors.

Conditioning can take many forms, depending upon which grain is being conditioned, but it always involves upgrading quality. Drying is probably the most common conditioning service and can be performed on all grains. Drying is performed to conserve grain with excess moisture. If drying were not performed in some crop years, a substantial quantity of grain might spoil, since high-moisture grain is difficult to store. Drying requires a considerable skill, because the application of heat to grain may cause factors other than moisture to be adversely affected. Once again the terminal merchandiser must be able to quickly tell where high-moisture grain will fit into the general merchandising program, taking into account how much moisture can be safely removed, and whether the grain, after drying, can be blended with other stocks in the elevator. The cost of the blending must, of course, also be considered.

Other conditioning services performed include clipping to raise the test weight of oats and barley; bleaching, to improve the color of oats; cleaning, to remove any unwanted material from grain; and washing or scouring, to remove fungus. Grain to be conditioned must be bought with sufficient discounts to reimburse the firm for the cost involved in performing these various services and, if possible, to return a profit.

Hedging Techniques of the Terminal Operator

Elevator operators are typically short hedgers, i.e. cash grain purchases are hedged by sales of futures. These sales usually take place within a short time after the cash purchase. The delivery month chosen for the futures hedge may be the current delivery month, although a better carrying charge may be earned by selling a more deferred month; if there is a possibility of an inversion in the nearby month, then a more deferred month will be chosen.

Initial placement of a hedge in one month does not mean that it must stay there. Changed or changing market conditions may cause the hedge to be switched to either a more current month or rolled forward to a more deferred month. Short hedges are usually moved to avoid making a delivery against the futures, to earn a better carrying charge, or to avoid an inversion.

If no delivery is to be made against the short futures position in a current delivery month, then sometime before the expiration of the future, the hedge must be moved out, rolled forward, to a later month. This can be accomplished by a spread order, i.e., an order to buy the nearby month (thereby closing out the position in that month) and sell a deferred month (thereby establishing a new position) at either a specific difference or the prevailing market difference at the time the order is placed. The difference sets the share or part of the full carrying charge that will be earned for the period. For example, if the short hedges are in the December and are to be moved to the March at 5 cents premium, then regardless of what the full carrying charge might be, the hedger has locked in 5 cents toward his costs of storage, insurance, and interest.

Sometimes, another futures market may be chosen for the initial hedge sale. This occurs when the local futures market is depressed relative to another market. As an example, a terminal in Kansas City might sell wheat hedges in Chicago futures if the Kansas City futures are depressed compared to Chicago. Kansas City futures are usually under pressure from hedge sales earlier in the crop year than Chicago, because the harvest occurs earlier in areas that usually hedge in Kansas City. A Kansas City terminal may hedge in Chicago late in the harvest, when Kansas City futures are under severe pressure; and then when Chicago futures are under heavy hedging pressure, the Chicago position may be liquidated and replaced in the Kansas City market. This can significantly improve the buying basis. The amount of the basis improvement would be the difference in the spread between the Chicago and the Kansas City futures at the time of the replacement hedge at Kansas City.

Another hedging technique is the *prehedge* or anticipatory hedge. If an elevator operator feels that large cash purchases may be made overnight, this volume is estimated and an equivalent amount of futures is sold before the close of the market to protect the buying basis. If this technique were not employed and overnight cash purchases were large, the pressure of hedge sales might force the futures well below the previous day's closing price, thereby worsening the buying basis.

If prehedging is employed and the volume of cash purchases is less than expected, however, the elevator is overhedged. In that case, the hedger will either buy in the excess futures sold or stay short (overhedged), figuring enough cash purchases will be made during the day to even up the position. If the futures the following day are lower than the prehedge sale price, the buying basis on later cash purchases is improved since it would be predicated on the prevailing market price.

Actual hedging decisions may be made and carried out by the merchandiser in charge of the particular grain, although in some firms these decisions may be entrusted to senior management whose principal duties are to supervise the hedging program. In this latter case, coordination and cooperation between the merchandising and hedging personnel is essential.

Terminal Buying Techniques

Elevators buy grain on to-arrive contracts as well as on spot. To-arrive bids are figured daily by terminal buyers within the framework of prior basis analysis and company policy governing it, competitive elevator bids, and bids received from users of grain. Changes may occur at any time. Early in the season, bids may be based on profitable delivery against the futures in addition to demand.

Overnight to-arrive bids are usually posted on the commodity exchange floor and bid direct into the country by the firm daily after the close of the market. Full information regarding price, grades and their premiums and discounts, time of shipment, means of delivery, and freight basis, if applicable, are set down. The merchandiser disseminates these bids by telephone or wire to subterminals and country elevators as well as to cash merchants outside the terminal market.

Grain may be bought as the bids are given out or, ordinarily, any time prior to the opening of the market the following day and during that trading session. Nearly all grain purchases made by the terminal buyer are made by verbal agreement over a telephone or face to face with the seller. From these agreements, the merchandiser writes up a booking giving full particulars on the trade. This booking is later typed into contractual form and mailed or presented to the seller. A booking and later a contract are prepared for spot purchases that involve a specific lot of grain in a particular freight car or truck.

Terminal Selling Techniques

Merchandisers are in constant contact with users of grain by telephone or wire. A few elevators use cash merchants or brokers when selling grain to allow more time for direct solicitation of prime accounts.

Customers of terminal elevators sometimes require grain with certain specifications other than those embraced by the federal grading system. For example, flour millers may ask for samples of wheat to determine how well it performs in actual baking tests. Oats buyers are interested in the lightness of the hull and the plumpness, brightness, and sweetness of the kernel. Distillers may pay a premium for plump rye, which has a maximum of 5% thin kernels. Maltsters select barley for its mellowness, which distinguishes malting barley from feed barley. Soybean processors select beans for their oil and protein content. Processors may also specify grain from a certain origin or a certain crop.

Offering prices for grain may be determined by the original buying basis or by the current basis cost of replacement, by competitive offerings from other elevators, blending or conditioning considerations, time of delivery, supply and demand, the cost of billing cancellation if applicable, and any pickup on billing switches that might arise on a particular sale. As with terminal purchases, most terminal sales are made verbally by telephone or face to face. After a trade is made the merchandiser writes up a booking giving full particulars regarding the sale. The booking is then typed into contractual form and either mailed or presented to the buyer.

Booking Freight

Before a shipper can quote a price for delivery to a destination by boat, he must know the pertinent vessel rates, the size and type of vessel needed, and the charter terms. A terminal sale often hinges on booking freight at a competitive rate for the particular time of shipment required by the buyer. Those terminal companies that do not own their own vessels employ the services of a vessel broker. Vessel brokers operate on a refusal basis, i.e., vessel space is offered to the elevator operator at a stated rate per long ton. This offering may be good from a few minutes to one day. Except for barge rates, vessel rates are not subject to ICC regulation, and therefore vessel operators are at liberty to charge whatever competition will allow. When freight is booked, a standard contract is drawn between the terminal operator or his agent and the vessel owner or his representative.

When a shipment is made by water, the carrier does not insure the grain, and it is therefore the shipper's responsibility to insure the cargo against marine perils. Usually the shipper also takes outturn insurance, which guarantees that the same amount of grain will be unloaded as was loaded into the vessel. The Great Lakes and the St. Lawrence Seaway are ordinarily closed from early December until mid-April, and insurance rates advance on the first day of December, with subsequent advances until all shipments have ceased.

Terminal Management Records

Elevator companies must keep three very important records in addition to employing many of the same systems and procedures used by the business community as a whole. These records are the long-and-short-position record, a billing record, and the elevator-stocks report.

The long-and-short-position record is the means used to record the trading position of the firm. A new long-and-short record is started each day, and a separate record is kept for each grain. A billing record may be prepared every two or three days by the traffic department of the firm for the use of the merchandisers, although it is usually a continuing record. The billing record provides an enumeration of types of inbound billing owned and, in the case of rail, may indicate possible destinations. The merchandising staff relies on this in replacing billing as well as in making sales. The elevator-stocks report enumerates by grain the grades, quality, and quantities in the elevator. It is vital to the merchandising staff in both buying and selling.

11

grain exporting

Directly or indirectly, anyone involved in merchandising grain is involved in the export movement of grain. The movement of grain to market, the use of the futures market for hedging and price reference, and the need to satisfy an export contract by accumulating and delivering an agreed-upon quantity and quality of a commodity at a specified place within stated time limits—these are all necessary elements in exporting grain as well as in domestic sales of grain. As the commodity moves through the total marketing route, these functions are carried out by grain producers, country elevators, subterminal elevators, terminal elevators, ocean-freight brokers, and final buyers.

While there are many similarities between exporting grain and the domestic trading of grain, several factors of varying complexity exist in the export process that do not exist in domestic grain trade. Governments throughout the world attempt to protect and subsidize their agriculture more than any other area of economic activity. Were it not for government restrictions, exporting of grain would be very similar to domestic trading except for the size of the transaction, financing considerations, and the long transportation hauls involved. There is also a much longer lead time needed in accumulating stocks of grain for export, and, generally, because of the increased risks of large quantities and long lead time, exporters usually find it essential to hedge their sales in futures contracts.

Growth of U.S. Grain Exports

The U.S. has been an exporter of grains since the middle 1800s. However, it was not until post-World War II that the U.S. began to develop a leading position in world trade in agricultural commodities. Postwar population growth and economic recovery and expansion in the U.S., Japan, and Western Europe led to large increases in the demand for food, particularly for meat protein. From the 1960s on, the U.S. and other large grain producers including Canada, Australia, and France, moved increasing amounts of surplus grain production into the world market.

In the 1970s, various events focused great attention on the significance of international trade in basic food and feed commodities—wheat, the feed grains, and oilseeds in particular. The effort to expand food production, with special emphasis on increased production of meat, was under way in the USSR and the People's Republic of China as well as in the U.S., Japan, and Europe. In 1971, the U.S. made the first of two devaluations of the dollar. This and the second devaluation which followed in 1973 had the immediate effect of making U.S. grain imports less expensive for foreign buyers, who became aggressive bidders. In the late 1960s and early 1970s, previously large government-owned surpluses of grain began to diminish under the pressure of heavy demand from foreign customers. Perhaps the single most critical event of the 1972-1973 crop production year was the very poor yield of the Russian wheat crop.

This created a sudden and unforeseen new source of demand for U.S. grain. Russia which had been a large net exporter of grain for long periods in its history suddenly became a major importer. China too was in the market for U.S. grain and oilseed commodities during the same period.

Even before these events occurred, there were many persons making doom-filled forecasts of starvation based on projections of population outdistancing the ability of the world's nations to feed themselves. The events of 1973 sharpened this debate significantly, embroiling consumer groups, politicians, and governments. The results included short-term meat boycotts and an export embargo on soybeans in the U.S. There were calls for the establishment of fixed grain reserves by many nations. Discussions on establishment of international grain reserves continue.

It is unlikely that U.S. consumers will soon return to the complacency they once had about the significance of U.S. agriculture in either the domestic or world economy. The events of the early 1970s have pointed up the essential nature of U.S. agricultural production in the world food situation.

U.S. exports of grain fall into one of two general categories, *dollar exports,* sometimes referred to as commercial sales, and *government exports,* which involve a variety of transactions conducted under government programs.

Commercial sales of U.S. grains for U.S. dollars or for barter have been the traditional form of export sales. Commercial sales may be transacted between U.S. exporting firms and foreign firms, or between U.S. exporters and foreign governments, with financing arrangements made between the parties without U.S. government involvement. Commercial exports for U.S. dollars generally account for about 20% of total U.S. exports.

An Example

Although there are many variations in both commercial (dollar) export transactions and in government exports, the following example illustrates the many complex considerations involved in all export transactions.

Assume a hypothetical firm, USA Grain, which has branch offices in several locations in the U.S. and a network of offices in foreign countries. USA Grain offers grain daily based upon its analyses of the market cost of accumulating the grain, the appropriate freight to get it to the various seaboards, the cost of *elevating,* loading the grain out at export elevators it may or may not own, ocean freight market, and all the ancillary charges of interest, outturn insurance, loading and unloading, stevedoring, and voyage insurance. It is necessary that USA Grain be constantly aware of all the cost and risk factors necessary to allow it to make a rapid decision when buying interest emerges. In the extremely competitive export business, the ability to move quickly often spells the difference between doing business or not.

In this instance, USA Grain receives a flat-price bid from a feed manufacturer in the United Kingdom for a 15,000 metric ton cargo, (590,000 bushels) 5% more or less at seller's option, of No. 3 Yellow corn for shipment the following month, *c.i.f.* (cost-insurance-freight) Liverpool, with import levy paid. Determining that it can cover such a sale profitably, USA Grain accepts the bid at 7:00 a.m. Chicago time (1:00 p.m. Liverpool).

USA Grain is now exposed to the risk of price change, plus the risk of a change in the levy, as of the opening of the Chicago futures market on 15,000 metric tons of corn. A price increase of even one cent per bushel will add $5,900 in costs for USA Grain. In the course of the previous business day, USA Grain may have bought corn stocks through its various branches which could be applied to this export sale. We will assume that 100,000 bushels were purchased the previous day that can be applied to this sale. This leaves USA Grain with the net risk position of 490,000 bushels as of the opening of the Chicago corn market at 9:30 a.m.

From this time until it has completely covered the sale, USA Grain has the following tasks to perform:

1. Covering in futures and/or cash purchases 490,000 bushels to limit its exposure to a rise in price, or a rise in the import levy, above the level upon which its quote was based

2. Surveying the ocean freight market for the proper size vessel in the required shipment period, and determining the loading range most advantageous (Atlantic, Gulf, Lakes, or St. Lawrence River) from which to fill the sale, by comparison of the f.o.b. price equivalents and available freight; the firm faces the risk of availability and increased costs on interior barge and, possibly, also lake vessels

3. Buying and moving cash corn stocks to the seaboard from which the shipment will be made

4. Lifting its hedges with offsetting sales of futures contracts as it makes the series of cash corn purchases

USA Grain would buy corn from country elevators through its own branch offices, and possibly from other cash grain merchants, to be delivered to the particular seaport through which the shipment will be made. In this instance, USA Grain would probably utilize all the segments of the grain marketing machinery to affect the completion of its contract.

In making the sale, USA Grain is risking misjudgment of likely price movement in the futures, changes in the basis, changes in the ocean freight, and, possibly, changes in the interior barge and lake vessel freight markets. It also faces the challenging logistics and timing involved in getting the entire cargo into position. For these reasons, a particular sale like this cannot normally be cut out from the firm's total, continual exposure in the export market, but must be considered as part of an ongoing exposure in both cash and futures markets. In light of this exposure, the firm is constantly involved in the ocean-freight market and in the minute-to-minute give and take of the total grain market. The matter becomes even more complex when, as in the previous example, to these considerations are added the elements of currency fluctuations (payment in pounds sterling instead of dollars, for instance), possible subsidy considerations, government programs involving different kinds of financing, and levies and duties applied in the importing country.

Role of Governments in International Grain Trade

Although one might expect that the lower the price of U.S. grain, the greater the volume of U.S. grain exports, this is not always the case. Government protection of agriculture in many importing countries often has the effect of making U.S. grain less price-competitive. The production of grain is stimulated by government protectionism in almost every country in the world where there is sufficient land to produce significant amounts of grain. The protectionism exists in both importing and exporting countries, and varying methods are used. The U.S. finds its largest competitors in the feed-grain export market to be Argentina, Canada, South Africa, and Australia, although many other countries compete on specific grains to certain destinations. These include Thailand on corn; Brazil and China on soybeans; and, at times, the USSR and other Eastern European countries on wheat, barley, and some oilseeds; and France on wheat, barley, and corn. Surplus-producing countries have an extremely large stake in the export of grains and will, either through direct subsidization of exports, long-term financing, or other methods utilizing the resources of the central government, compete vigorously for the markets available to them, regardless of price.

Although the Common Market countries are net importers of grains, through the imposition of high internal price supports and protective levies on imports, they have stimulated production of grains and related food products in their efforts to attain agricultural self-sufficiency. Because of the levy system in the Common Market, U.S. corn (and other foreign corn) is not allowed to trade in the interior of these countries at less than the arbitrarily established target price. This has, at times, been more than twice the price at which imported corn would have traded were there no levy added on at the point of entry.

Japan has become the best market for U.S. exports because of the rapidly increasing standard of living in that country and the fact that grain enters Japan virtually without import duty. Thus, U.S. grain is able to compete in the Japanese market on a free-market basis with the production of other countries.

		U.S.		
		production	**exports**	
year	**commodity**	(millions of bushels)	(millions of bushels)	(% of total production)
1940	**wheat**	815	40.6	4.9
	corn	2,206	14.8	.06
	soybeans	78	.085	.10
1950	**wheat**	1,019	374.0	36.7
	corn	2,764	109.7	3.9
	soybeans	299	26.9	8.9
1960	**wheat**	1,355	631.1	46.5
	corn	3,907	276.5	7.0
	soybeans	555	143.2	25.8
1970	**wheat**	1,370	715.5	52.2
	corn	4,099	506.1	12.3
	soybeans	1,124	424.2	37.7
1976*	**wheat**	2,212	1,100	49.7
	corn	6,216	1,650	26.5
	soybeans	1,264	690	54.5

*estimated

Role of U.S. Government

The significance of the U.S. government policies in grain exports are felt in three areas. The first of these is the overall agricultural program which in recent years has stimulated full production of wheat and the feed grains in order to provide for export demand. The government influence is also felt in the subsidy program for U.S. exporting firms which better enables them to compete in foreign markets. The most important government influence on grain exports began in 1954 with the passage of the Agricultural Trade Development and Assistance Act, better known as Public Law 480. And finally, government actions taken thus far in the 1970s to restrict or limit export of grains in limited supplies, have had very significant influence on U.S. grain exports and their competitive posture in world markets.

Background

In the 1950s, government support programs and growing agricultural production created surpluses to be disposed of by the U.S. government. Year after year, American farmers were producing more wheat and feed grains than the domestic market could absorb at the price levels then in effect. After a period in which the Commodity Credit Corporation handled most of the country's exports for dollars on a direct-sales basis, a program was devised whereby the private exporters would handle the bulk of the exports and would be paid subsidies established daily by the USDA. These subsidies were necessary to keep the price of U.S. grain competitive with grain from other sources.

Such a subsidy program still exists, although it is relatively less important than in earlier years. When subsidies are in effect, they are established daily by the USDA in Washington on the basis of information available relative to the world market price of competing grains. Prior to the actual shipment of grain, the U.S. exporter must register his sale with the government to collect the subsidy payment that is necessary to equate his sale to one of a strictly domestic nature.

Public Law 480

Initially PL 480 was regarded as a short-term program to help move growing surpluses of grain to foreign markets. It had the added purposes of helping the economic development of foreign nations, enlarging the traditional dollar markets for U.S. grains, and providing emergency relief for disaster-stricken nations. The act has been amended and extended several times and is no longer regarded as a short-term program. It has become a uniquely successful program in boosting U.S. exports of grain.

PL 480 provides for a variety of export transactions under five titles of the act. Title I provides for government-to-government export transactions with financing on either short-term or long-term basis in the foreign currency. Between 1956 and 1967, grain exports under Title I accounted for more than 70% of all PL 480 exports. In recent years, however, only 18% to 20% of PL 480 exports have been conducted under Title I. In this period, a large amount of the PL 480 exports has been shifted to the Title IV category as transactions with long-term financing in U.S. dollars.

Title II involves government-to-government donations of U.S. agricultural commodities in emergency and disaster situations, and for economic development. In the past 10 years, such donations have ranged from 3% to 30% of all PL 480 exports.

Title III shipments of grain represent transactions involving government-to-government barter for commodities strategic to the U.S., and donations made by U.S. charitable organizations from CCC stocks of grain. During the past 10 years, Title III exports have ranged from 13% to 18% of PL 480 exports.

Title IV provides for long-term credit arrangements in U.S. dollars between the U.S. and friendly governments. In most recent years, Title IV exports have amounted to roughly 40% of all PL 480 exports.

While PL 480 exports have grown since the program was implemented, they have not grown at the expense of traditional commercial dollar exports. PL 480 shipments represented added business that would not be conducted were it not for the program. Since the 1950s, commercial dollar sales of U.S. grains and oilseeds have also grown significantly with the increasing economic strength of foreign nations, which in part has been aided by the PL 480 program.

Grain Exports and the U.S. Economy

The important role of U.S. agricultural commodities in meeting world food needs is widely recognized, even resented. What is less apparent to both U.S. consumers and some foreign nations is the unparalleled efficiency of American agriculture, which enables the U.S. to provide for its own and a significant amount of the rest of the world's food needs. In the U.S. today, one man-hour of farm labor produces nearly nine times more output than did the same man-hour in 1919. Viewed in another way, today's farmers produce 53% more food, feed, and fiber on 6% fewer acres than were utilized a generation ago. These advances are in no way accidental. They are the product of scientifically selective use of land, plant, chemical, labor, mechanization, and capital investment inputs, which are available to individual enterprise in few other economies of the world.

In 1975, the value of U.S. agricultural production was estimated at $99.2 billion, greater in value than the output of any other industry. Agriculture also led all other American industries in the value of its exports. The value of 1976 agricultural exports, with grains and oilseeds accounting for approximately half, has been forecast at $22 billion, roughly a fifth the value of all U.S. exports. In the past several years, agricultural exports have been the determinant factor in keeping the U.S. balance of payments position in surplus condition.

12

grain purchasing by processors

The steps in the marketing of domestic cash grain are completed with the purchases of grain by processors. Wheat, corn, oats, rye, barley, and soybeans are each used in the manufacture of the numerous grain products. Most of the crop of some grains is processed; in other grains, the bulk of the crop is used for livestock feeding and only a small portion is processed. A processor purchases the type, variety, and grade of grain best adapted for manufacturing his particular product, and, whenever possible, hedges his purchase of grain and sales of grain products with offsetting, opposite sales and purchases of futures contracts.

Usage of Grain by Processors

The more important processors of grain are flour millers, wet processors of corn, dry processors of corn, oats millers, maltsters, soybean processors, distillers, and mixed-feed manufacturers.

During the past five years, exports of wheat have been the largest area of consumption, averaging nearly 50% of total U.S. wheat supplies. In most years, domestic food use of wheat in flours and bakery goods has been the second largest area of consumption ranging from 21% to 26% of total U.S. wheat supplies. About 3% to 5% of wheat supplies are used as seed for future crops, and 3% to 8% of wheat supplies are used as animal feeds. Typically, 60% to 65% of the flour milled in the U.S. is consumed domestically by the commercial baking industry.

The principal grains used by flour millers are wheat and rye, and their major finished product is flour. Inasmuch as rye is not used very extensively as livestock feed, most of the U.S. rye crop is utilized commercially in the manufacture of rye flour, rye whiskey, and rye malt.

In the period 1971-76, from 65% to 75% of the corn produced in the U.S. was utilized for livestock feeding. Approximately 3% of the corn crop grown in 1975 was used by wet corn processors, 2% by dry corn processors, 3% by distillers, and 3% by manufacturers of cereals. About 62% of the corn supply was used for animal feeding, and exports consumed 27%.

Wet processors of corn use shelled corn in the manufacture of a diversified group of products including starch, dextrin, corn syrup, corn sugar, corn oil, and feeds. Dry processors of corn use yellow shelled corn to produce granulated corn meal, bolted (sifted) corn meal, brewers' grits, hominy grits, and flour. They use white shelled corn in the manufacture of pearl hominy, brewers' grits, hominy grits, granulated white corn meal, bolted white corn meal, and white corn flour.

More than 80% of the annual U.S. oat crop is consumed as livestock feed. The principal products of oats millers are rolled oats, steel-cut oatmeal, and ground oatmeal. These milled products usually consume 10% to 14% of oats supplies. In recent years, exports of oats have ranged from 5% to 10% of U.S. supplies.

Roughly 40% of the 1975 U.S. barley supply was used in the malting industry, a small additional percentage is normally used in the manufacture of pearl barley. Maltsters use barley in the preparation of malt, a grain product whose starch content has been converted into a form of sugar through the process of sprouting or growing under controlled conditions of humidity and temperature. Some rye as well as a small quantity of wheat is used in the preparation of special-purpose malt products. In 1975, 48% of U.S. barley production was used as animal feed and 7.8% of the crop was exported.

In recent years, from 45% to 55% of the U.S. soybean crop has been processed for crude soybean oil and soybean meal. In the processing year 1974-1975, it was estimated that 825 million bushels of soybeans were crushed yielding 18.8 million tons of soybean meal and 8.6 billion pounds of crude soybean oil. In the same period, exports of soybeans accounted for 30.7% of total supplies, 4.4% was used for feed and seed purposes, and it was estimated that 16.4% of total supplies would be carried over for use in the following processing year.

The percentage of the various types of grain used in manufactured feed products varies considerably. Corn, sorghum, oats, barley, and wheat are used in the preparation of such mixed feeds as poultry mashes, scratch grains, horse feeds, hog feeds, dairy feeds, and pet foods. Corn, rye, and barley malt are the principal grains employed by the distiller in the manufacture of bourbon, rye, corn whiskey, and high-proof alcohol or neutral spirits.

Purchasing Methods of Processors

Different types of processing industries employ various methods in purchasing needed grain supplies, and varying methods are used even within a given industry. A small amount of grain is purchased directly from farmers, and some grain is bought directly from country elevators. Processing industries tend to purchase the majority of their grain needs from terminal market elevator companies.

Some processors have plants located in the surplus grain-growing areas and they are able to draw most of their grain requirements from the immediate area. The larger processors, who need large quantities of grain and have additional plants located in manufacturing areas, tend to draw their supplies from considerable distances and usually prefer to purchase the bulk of these supplies in the terminal markets. Perhaps the most important advantage of purchasing grain in the terminal markets is that these markets are the channels through which grain from a large surrounding area passes and the terminal market elevators can usually offer a large amount of grain either on spot or out of terminal elevators at any time the processor wishes to make a purchase. The processor is also better able to select his grain in the terminal market because of the wide variety of types available to him.

Some of the larger processors retain their own buyers in the terminal markets to purchase grain from cash grain commission merchants or terminal elevators, wherever the most advantageous price can be obtained. Other processors purchase supplies in terminal markets through brokers, while a third group of processors buy solely from terminal elevators or grain merchandisers and do not have their own representatives in the market.

Processors usually do their heaviest buying at the time crops are moving from the farm into storage. At this time, a wider selection of types and grades of grain is obtainable, making it possible for the processor to obtain the qualities of grain desired, and the cash basis is usually attractive for buying. The extent to which a processor is able to purchase supplies at this time is limited somewhat by his storage capacity, although a processor may rent additional storage space.

Most processors buy grain in the market throughout the entire year, but because of certain production peculiarities this may not be true for all processing industries. For instance, in the malting industry it is the custom to close plants about June 1 for the summer months. Maltsters buy heavily in the fall, but they taper off in the spring after March purchases.

The biggest problem in the selection of grain by the processor is that of obtaining the type and grade of grain required for his particular product. Numerous guides exist to aid the processor in selecting his grain, but the extent to which these are used depends on how exacting his requirements are. The first guide to the processor in the purchase of grain is the federal grading system, which is relied on solely by certain processors in purchasing their raw material. Another aid to the processor in purchasing grain is appraisal by an experienced buyer. By examining the grain on the basis of its general appearance and soundness, the experienced buyer is able to determine certain qualities of the grain. By such examination, grading factors are checked and other characteristics are determined which might not be wholly indicated by the numerical grade assigned by the licensed grain inspector. Further assistance is given to the processor by tests for qualities not included in the federal grain standards. In some cases, tests are made under conditions similar to those existing when the product is actually processed so that the processor is able to determine how a particular lot of grain will react when it is milled or processed.

Use of Futures Market by Processors

In buying grain and selling products, processors sometimes have large quantities of grain on hand not covered by sales of products, and, at other times, they have sales of products not covered by purchases of grain. Under these conditions, the processor is subject to loss through fluctuations in the price of grain. Since the processor has made an investment in his plant for the purpose of securing a reasonable return for services performed in processing and merchandising grain products, he is desirous of handling his grain and products inventories in such a way as to avoid all possible price risk. Although the simplest way to accomplish this is to buy grain as the products are sold, this is seldom possible, because the processor must buy grain when it is available and sell his products when he can.

The processor surmounts this difficulty by hedging in the futures market. In order to protect themselves, flour millers, for example, must buy wheat in sufficient quantity at the time the crop is moving in heaviest volume. This period generally affords them a better possibility of obtaining choice wheat without paying excessive premiums. In buying grain in advance of flour sales, a flour miller subjects himself to a possible loss resulting from fluctuations in wheat prices. He guards against such losses by simultaneously selling futures as the cash purchases are made. This hedged position is maintained until flour is sold against wheat on hand, at which time the hedge is removed on an equivalent amount of wheat.

The flour miller is also faced with the necessity of selling flour ahead as much as 90 to 100 days. In many cases, he does not have the wheat on hand to fulfill such commitments, and it may be inadvisable to buy the wheat at that time; perhaps he does not have the storage space even if the wheat were available at an attractive basis. If he wants to keep his customer, however, he must sell flour when the customer wants to buy. In such cases, when he has priced flour for several months ahead and has no wheat with which to grind the flour, the flour miller hedges his position by simultaneously purchasing wheat futures as flour sales are made. Later when the cash grain is purchased to make the flour, the hedge is simultaneously removed by selling wheat futures or by an exchange of such futures for cash wheat with the seller of the grain (see Chapter 4).

In hedging operations, the flour miller hedges his net position, which means that stocks of grain on hand are balanced against sales of flour. If the former is greater than the latter, his position in the futures market will be short to make up the difference, and if the contrary is true, his position in the futures market will be long to make up the difference.

It is important for the processor not only that cash grain prices move with futures prices, but also that the price of grain products shows a correlation to grain prices. If product prices do not vary with grain prices, then the value of hedging is reduced. The flour miller is fortunate in this respect because the price of flour (his principal product) is based on wheat futures prices, and whenever the wheat futures price changes the flour miller adjusts his flour price.

In calculating flour prices, the flour miller takes the wheat price and multiplies it by the number of bushels of wheat required to produce a sack of flour in his plant. To this figure (known as the *conversion rate*), he adds an amount to cover milling costs, selling costs, and profit, a differential depending on the grade of wheat used, and a charge for the sacks. From the resulting total, he deducts the value of the by-products to arrive at the price of a sack of flour.

While hedging protects the flour miller in large part because of the relationship between flour prices and wheat prices, the miller is not protected on the by-product portion of his production. Since on an average it takes 2.3 bushels of wheat to make a sack of flour, about 38 pounds of millfeeds are produced with every sack of flour. This by-product is used as animal feed, and its price is dependent on the price of feed rather than on the price of wheat. Feed prices are partially independent of wheat prices because they are influenced by such factors as available pasture lands, livestock population, season of the year, and competition from animal feeds of other kinds. Thus, approximately one-third of a flour mill's purchases are converted into a product largely unaffected directly by wheat prices. Obviously any appreciable variation in the recovery value of one-third of the raw material will definitely affect the profit or loss of the miller.

In addition to the risk connected with the sale of feed, the processor of flour also has the same risk as the elevator in the matter of changes in basis on his cash wheat. In fact, the flour mill's risk in this respect is probably accentuated because the flour miller buys considerable premium grain that he cannot deliver on futures contracts except at a loss.

In addition to the use of the futures market in limiting wheat inventory losses, the flour miller on occasion also uses the futures market as a means of securing supplies. Sometimes, taking delivery of wheat against futures will enable a mill to buy its requirements more cheaply than would cash wheat purchases. Such a delivery of wheat on the futures contract can be simultaneously hedged by a corresponding sale of a deferred future. For example, a delivery of Chicago December wheat might be hedged by a sale of an equivalent amount of Chicago May wheat.

In its hedging operations, the flour mill has the same problems as the terminal elevator in deciding in which month to place its hedge and also in selecting the market in which the hedge is to be placed. The flour miller is also able to make profits by watching his position carefully and switching hedges when opportunities arise.

While these examples have referred primarily to the flour miller, the same principle of hedging in the futures market also applies to other grain processors. For the distiller, the breakfast-food manufacturer, or any of the various industries that utilize grain or its products, the principle is the same. Hedging affords a maximum of protection for both producers and consumers and correspondingly reduces the margin of profit necessary to cover the element of risk. This factor in turn makes for a minimum of price differential between original producer and final consumer and, to that extent, lowers costs.

13 *flour milling*

Wheat flour has been used for human food as far back as recorded history. Early methods of production included the mortar and pestle, the saddlestone, and a rotary-motion device called a *quern*. *Milling,* the removal of the flour from the other constituents, developed out of the necessity to break down the structure of the wheat kernel, which is made up of three principal parts: (1) *bran,* the outer covering; (2) the inside *endosperm* of gluten and starch particles; and (3) the *germ,* the embryo wheat plant. Millstones and water mills were in use in ancient Greece and Rome, and in northern and western Europe at the time of Julius Caesar. Windmills appeared in Europe in the 12th century.

History of U.S. Flour Milling

The Plymouth settlers planted English wheat they had brought with them to North America, and by 1660 the New England colonies had a surplus of the grain. The first mill in the U.S. was erected in Boston in 1632. This first effort was a wind-driver type, but the more-efficient mills driven by tide power soon became common. These were followed by mills erected in *millraces* (artificial canals). As technical skill increased, mills were built on streams and inland rivers. By 1700, most New England villages had a gristmill of some kind, a few of which survive to this day.

The Connecticut Valley was the first region to develop a large surplus of wheat, but wasteful methods soon exhausted its soil, and, by 1800, this region was importing its wheat and flour. The flats of Escopus, in New York State, became the next large wheat-production area, followed by the Hudson Valley, the Mohawk Valley, Vermont, Virginia, Maryland, and Pennsylvania.

Philadelphia became the first important milling center because of its waterpower and the water transportation available from existing wheat-growing areas. It was there that the first all-mechanical flour mills were built, which visiting foreign millers proclaimed to have a higher degree of mechanical excellence than those of Europe. Conveyers and elevators used in these mills for transferring wheat or its products from one machine to another were unknown in Europe at that time.

Richmond, Virginia, soon followed as a milling center, and it was here that the first mill with a capacity of 900 barrels a day (known as the Gallego Mill) was built. No other mill exceeded this capacity until General Washburn built his Minneapolis mills many years later. By 1840, Baltimore took the leadership as the foremost milling center and held it for some time, aided by the construction of railroads that made wheat available from the most favorable existing wheat-growing areas. By 1859, Baltimore had 59 mills.

Simultaneous with the development of milling in Baltimore, new wheat areas developed in the valley of the Genessee in western New York, in lower Michigan, and in Ohio, causing the erection of many small mills in these territories. Rochester, located on the Genessee River, with splendid waterpower and water transportation on the Genessee Canal to the south, the Erie Canal to the west, and Lake Ontario to the north, soon developed into a milling center. With completion of the Erie Canal to take its products to the New York and export markets, Rochester became such a milling center that it was referred to as the Flour City, a distinction that, in later years, it lost to Minneapolis.

Oswego, New York, also having fine waterpower and access to Canadian wheat as well as to that of Illinois, Michigan, and Ohio, soon developed a large milling capacity. Its growth was assisted by a branch canal connecting with the Erie Canal at Syracuse, making available New York and export markets. Conditions of wheat production and transportation in this country were changing so rapidly at this time that Oswego's position as a milling center was soon lost.

Up to the time of the development of the northwestern spring wheat areas, most of the wheat produced in this country was of the soft winter variety, with vitreous varieties selling at that time at a discount. Wheat production was soon extended southwesterly until it reached the Mississippi River at St. Louis and a short distance beyond. The opening of this vast wheat-growing area was responsible for the erection of considerable milling capacity at St. Louis, where water transportation on the Mississippi River to New Orleans also made export markets available. The building of railroads through the wheat-growing sections brought about the erection of larger interior mills at points where rail rates were favorable for obtaining wheat and transporting flour to the consuming markets.

In the Northwest, where the production of soft wheat was not possible because of climatic conditions, the planting of spring wheat became a necessity. This variety was first produced in northern Illinois, Wisconsin, and southern Minnesota. As Chicago and Milwaukee were then the natural markets for that territory, the first mills to grind this wheat were built at these points.

Following 1851, the increasing Minnesota settlers from Iowa, Wisconsin, and Illinois were joined by immigrants, who were arriving in great numbers at that time. The movement soon extended west into South Dakota and North Dakota, resulting in a large productive area of spring wheat. Minneapolis became the natural market for this wheat, and its location on the Mississippi River made it possible for a large spring-wheat milling capacity to be established there.

About the time spring-wheat production in the Northwest was nearing its peak, settlers in the Southwest began producing a wheat called Hard Winter wheat. The first varieties grown in that section did not yield a satisfactory flour, so a new variety called Turkey Red was brought from Russia for seed. This proved to be a better wheat for the climate in that section and produced a flour quite suitable for bread. The production of this wheat (Hard Red Winter wheat) now extends over such a large area that it exceeds the production of spring wheat. A large portion of the wheat produced in the Southwest is exported, however, largely because of ample supplies and the short haul to ocean transportation.

After settlers crossed the Rocky Mountains, a considerable amount of wheat was produced in the Pacific Northwest, mostly of a soft variety This production brought about the erection of considerable milling capacity in the Pacific Northwest with a large share of the flour production being exported to the Orient.

Selection of Wheat for Milling

The first and one of the most vital steps in milling process is the selection and accumulation of wheat. Although many smaller mills throughout the country do a largely local business, drawing their supplies of wheat from the adjacent territory and marketing their flour locally, the larger mills find it desirable to make most of their purchases in one or more of the large terminal markets. Perhaps the most important advantage of purchasing grain at one of the larger markets is that these markets are the natural channels through which wheat of the surrounding territory passes. A terminal market can offer a miller a large amount of wheat either on spot or out of a terminal elevator at any time he wishes to make his purchases. The buyer is also better able to select his wheat because of the wide variety of types offered to him. From the price standpoint, the terminal market also seems the best source of supply for the miller.

Flour millers probably go to greater lengths in selecting the type of wheat they desire than does almost any other type of grain processor. To produce a certain grade and quality of flour, the miller must have a specific type of wheat.

Generally speaking, a miller produces two types of flour: bread flour and cake flour. The bread flour may be further divided into bakers' flour and family flour. The bread flours are ground largely from Hard Red Spring and Hard Red Winter wheats, while the cake or pastry flours are usually ground from Soft Red Winter and White wheats. It is generally believed in the trade that the baking industry uses about 40% to 45% of the flour produced by mills, the family trade about 45%, and that the remaining percentage is absorbed by institutions such as hospitals, hotels, restaurants, and miscellaneous outlets.

Because of their high *gluten* (protein) content, bakers' flours are generally considered strong flours, which means that they produce a loaf of bread of good volume and texture under a variety of baking conditions. *Bakers' flour,* as the name implies, is used largely by commercial bakers. *Family flours* are generally lower in gluten than bakers' flours. This flour is intended to be an all-purpose product that will give reasonable satisfaction in ordinary home baking of cakes and pastry as well as bread.

Cake flours are used primarily in baking cakes, pastries, cookies, and crackers. This flour is usually ground from soft wheat of lower protein content. While a high-protein content is desirable in bread flours, a low-protein flour usually gives better results in cake baking.

In buying grain for all types of flours, the miller is constantly confronted with the problem of quality. Consumers of flour expect a given flour to perform in the same manner in the same formula at all times. This expectation makes it necessary for the miller to produce a flour that consistently has the same chemical and physical factors for a given grade.

Wheat selection is difficult because of the fact that the baking quality of wheats of like grade and appearance may vary widely from one to another,

Local soil and climatic conditions have much to do with quality, and two lots of wheat of the same variety may vary considerably in baking qualities. For instance, a sample of No. 2 Hard Red Winter wheat may bake an excellent loaf, while another sample with the same grading factors may bake an inferior loaf. Wheat grown in one locality may vary greatly from year to year in baking quality. Thus the miller must keep constantly alert for the best baking wheat available. Although from year to year he produces essentially the same types of flour, from year to year his supplies may be drawn from entirely different sections of the country and his mix composed of entirely different types of wheat. Sometimes millers must revise standards because of the quality of the new wheat crop, although bakers', family, and cake flours produced by one flour miller generally perform rather consistently year after year.

The factors involved in the federal grading of grain are of considerable importance to the flour miller. Grain must be considered on the basis of moisture content, soundness, mixtures of other grains, and test weight. Musty or otherwise unsound wheat is not suitable for milling purposes because it will produce a flour that will be musty or otherwise unsatisfactory. Smutty wheat, contaminated by fungus, may be prepared so it will be suitable for flour-milling purposes, provided that it is properly cleaned by washing or scouring. Smut and mixtures of other grains and seeds are objectionable unless they can be removed from the wheat before milling, because they tend to affect the color and, in some cases, the odor and keeping qualities of the flour.

A. Quality Factors

While grading factors are important to the flour miller in the selection of wheat, other considerations are often of even greater importance. Probably the most important single factor in purchasing wheat of milling grade is protein content. Wheat contains several proteins, but the two principal ones are *gliaden* and *glutenin.* Although several cereals contain gliaden, wheat is the only grain that has glutenin in appreciable quantities. When flour is moistened, gliaden and glutenin combine to make a substance called gluten, which has an elastic property. When a loaf of bread dough is prepared and baked, the gases expand but are prevented from escaping by the gluten; thus it is possible to obtain a light loaf of bread from wheat flour, whereas it is impossible to obtain a light loaf of bread with flour made from other cereals that do not have gluten with its elastic property.

In a general way, the strength of bread flour varies with the gluten content of such flours. Analysis of the protein content of flours and wheat has come to be used as the common basis for determination of gluten content, and thus indirectly as a measure of bread-baking quality. Since the flour protein content bears a parallel relationship to the wheat protein content, the use of protein tests has become quite general in the selection of wheat for milling various flours.

Another quality looked for in wheat, which is not considered in the federal grain standards, is ash content. Ash is the mineral matter that wheat has taken from the soil during its growth. Most of wheat is organic matter, which is combustible, but ash is the part that does not burn and is primarily phosphate. The ash test involves burning the flour and reweighing it after everything combustible has been consumed. This test is a good indication of the amount of flour that can be extracted; it is not always an indication of the quality of the flour.

Millers look for wheat that has excellent appearance, grades well, and has high protein content. Because mills sell flour for baking performance, that factor is primary. More and more importance has been attached to milling and baking tests which determine the potential flour recovery and the bread-baking qualities of a particular sample of wheat. The wheat grown in one area may vary widely from year to year from wheats grown in other areas because of varying climatic and soil conditions. The flour miller therefore, must know in advance of the movement of the crop where the type of wheat he needs for his flour may be obtained. Some of the larger flour mills send survey crews into the wheat-growing areas. The surveys are started in the South when the harvest begins and move northward as the harvest moves northward. Survey crews obtain hundreds of samples of wheat, which they forward to laboratories for milling and baking tests. The results of these tests are mapped and charted, and they are then used by buyers in purchasing wheat supplies. Smaller mills also study wheat territories, but on a less comprehensive scale.

B. Price Factors

One price factor considered by the wheat buyer is the *yield value* of the wheat, which is the percentage of flour that can be obtained from a given amount of wheat. This quality is usually referred to as the amount of wheat necessary to produce a sack of flour—100 pounds net. A yield value of 2 bushels and 15 pounds, for example, means that 2¼ bushels of this wheat would be needed to produce a sack of flour.

Yield value varies directly with the test weight per bushel of wheat and inversely with the moisture percentage in the grain. A sample of wheat weighing 60 pounds per bushel and testing 11% moisture will yield more flour than a sample testing 60 pounds and containing 14% moisture, just as a sample testing 55 pounds will yield less flour than a sample testing 60 pounds.

Experience may prove to a miller that 55-pound wheat at 5% per bushel discount under 60-pound wheat may be a more expensive purchase for him than 60-pound wheat. This would be true if the cost of the quantity of the 55-pound wheat necessary to produce a sack of flour were higher than that of the 60-pound wheat required for the same purpose. For instance, suppose 2½ bushels of 55-pound wheat or 2¼ bushels of 60-pound wheat will each produce a sack of flour. If the miller paid $2.00 per bushel for 60-pound wheat, the wheat for a sack of flour would cost $2.00 × 2¼, or $4.50; 55-pound wheat at 5% per bushel discount would cost $1.90 × 2½, or $4.75. Under these circumstances it is cheaper to buy 60-pound wheat even though its cost per bushel is 5 cents higher. Potential production of millfeed as a by-product of flour milling, and profits from the sale of the millfeed must also be taken into account.

Price must also be considered when choosing between types of wheat. If the same standard can be maintained by a blend of 50% Kansas Hard Red Winter wheat and 50% Central States Hard Red Winter wheat as by a blend of 25% Hard Red Spring wheat and 75% Central States Hard Red Winter wheat, the miller will probably be guided by the relative cost of the two mixtures. He may find the Spring-Central States blend is cheaper and will use it. At some later time, the Kansas-Central States blend may be cheaper, and he will switch to that mixture. Or perhaps he will discover that by using a greater percentage of strong Kansas or Spring wheat he will be able to include a cheaper quality wheat that will lessen the net cost of his mix without impairing the quality of his flour. Calculations of this kind must be checked constantly, and the miller must keep in touch with prices and quality of all wheat at all times if he is to grind the most economical wheat and keep his flour prices competitive.

Storage and Preparation for Grinding

After the wheat is selected, the miller must store his purchases until they are used. If the miller is to deliver a constant mix into the milling process, he cannot throw all of his wheat into one bin. Wheat must be separated according to protein, baking quality, type, yield value, or other considerations that may be desirable from time to time. With the wheat thus separated, it is a simple matter to blend these various lots together in the desired percentages. In most modern mills, this is done by mechanical percentage feeders that are attached to bins containing the component parts of the mix. Each feeder is set to deliver the desired percentage from its bin, and a master drive runs all the feeders as one. In this way the wheat is drawn together in the desired percentage and delivered to the cleaning machinery of the mill. As the various wheat bins run empty, they are refilled from the miller's stock on hand. This stock of wheat must be turned or aerated frequently to maintain its condition and must be watched constantly so that there is no spoilage as a result of fermentation or insect infestation.

Cleaning machinery used in mills varies in different parts of the country because of the special characteristics of the wheat or the impurities normally contained in it. For instance, in the western states, smut is quite common, and cleaning machinery that is especially designed to remove smut balls and spores of smut must be used. In certain sections of the southern and eastern soft-wheat-producing area, an admixture of garlic bulblets is a problem. After thoroughly drying the garlic bulblets present, much of this impurity can be cleaned or drawn from the wheat with air. Other machinery is especially designed to remove weed seeds, chaff, straw, and other impurities. Aspirators draw out the dust from the wheat by suction. The wheat is dry-scoured and, in some instances, washed.

In order to facilitate grinding and to make possible a cleaner separation of bran from flour, it is the general practice of mills to temper the wheat—to add a suitable amount of water to the wheat and allow the mixture to stand for several hours before grinding. The addition of water toughens the bran coat of the wheat kernel and causes it to separate more completely from the flour. The amount of water added and the time allowed to elapse before milling vary with the wheat used and its moisture content. For example, more water is added to a low-moisture wheat, and a hard vitreous-type wheat is allowed to stand longer. It is as important not to temper wheat too long as it is to be sure the wheat is tempered sufficiently. An overtempered wheat will make it difficult to separate the endosperm from the bran and will cause flour particles to adhere to the bran, thus impairing the normal yield. To remove any bran particles that may have been loosened during the process of tempering, it is customary to scour the wheat again before grinding.

Grinding

The tempered wheat is then fed to the mill proper. The first grinding is accomplished by a series of break rolls, which consist of pairs of corrugated iron rollers rotating toward each other, one at about 2½ times the speed of the other. Usually a battery of five or six sets of break rolls is used. The first break rolls are usually quite far apart, just cracking the wheat. The second break rolls are somewhat closer, and the others successively closer, until a fine separation of bran from flour is made.

After each break, the cracked and broken wheat is bolted by large sifting devices that separate the fine flour particles from the coarser bran. The coarser portion of each break is returned to the next set of break rolls for further reduction. This grinding and sifting process results in three general groups of products: the coarse fragments, which are reground until nothing but the bran remains; the fine particles, which are the flour; and an intermediate group, known as *middlings*. These middlings are further reduced by grinding between smooth rolls, recovering some additional flour and reducing the middlings until what is left is suitable only for feed.

The fine flour particles are graded and blended according to the flour desired; then they are bleached and packed under the brand of the milling company.

Grades of Flour

Flour falls into three classes or grades: straight, patent, and clear. A *straight flour* is all of the bolted (sifted) meal after all of the feeds such as bran, middlings, or shorts have been removed. A *patent flour* is the more refined portion of straight flour, and the residue is known as *clears*. The latter are further classified as first clear and second clear. The terms *short patent* or *fancy patent* are often used in flour advertising; they merely reflect the manufacturer's contention that his flour is more highly refined. For example, a 70% patent flour would imply that it consisted of the best 70% of the total. As a rule, clears are regarded in the U.S. as animal feeds, although they contain nothing harmful to humans. Clears from grinding rye flour are used in blending to make light and medium rye flour.

A mill usually packs several different grades of flour under various trade names. It is naturally expected by the consumer that these flours will perform in the same manner at all times. The larger bakers require a flour that will have the same chemical analysis at all times. This necessity for standardization requires the miller to keep a constant check on his products. All mills of any size maintain laboratories and a staff of chemists and technicians whose duty it is to test the flour streams several times daily for the chemical accuracy and baking quality of the products. This careful control of the quality of his product is the miller's assurance that his deliveries are in every way fulfilling his contract with the customers.

Merchandising Techniques

Mills generally employ one of two methods in merchandising their flour: either a mill has its own salesmen covering the territory in which it wishes to dispose of its product; or a mill secures the services of a flour brokerage concern, whose salesmen cover the different markets to which the mill has favorable transportation rates.

It is customary for the mill at the close of each market day to wire its representatives, whether salesmen or brokers, a price on its different products, basing such a price on the selling price at the close of that day's market for the grade and quality of wheat required to make the product quoted. Sales are generally made for delivery within a period of 120 days. Sales made for contemplated delivery beyond 120 days are usually subject to a specified storage charge to be paid by the buyer to the miller on the basis of so much per sack per day beyond the 120-day limit. This provision for the collection of storage charges is necessary to cover the expense of storing grain and to cover insurance and interest charges incurred through the ownership of grain carried as a protection against flour sales.

Sales made for actual delivery farther ahead than 120 days embody considerable risk to both buyer and seller because of the element of price fluctuations in the premium or discount levels of the cash grain (changes in the basis), and the possibility of substantial fluctuation in feed prices from those in effect at the time the flour sale is made. Because of the risks, the milling industry has attempted to concentrate its selling activities on the basis of making deliveries within the 120-day period from the date the sale is consummated.

In order to protect his quality program, it is often advisable for a miller to accumulate a stock of wheat and to carry it for several months before grinding it. Millers customarily protect themselves from investment losses during the period when they have grain in storage, either by selling flour for delivery in the future at the same time the wheat is acquired or by hedging in the wheat futures market.

14 wet milling of corn

Corn is native to the Americas. Columbus discovered it in Cuba, and other early explorers found it growing along the Atlantic seaboard as far north as the St. Lawrence River. The word *corn* was applied to the newly discovered cereal by explorers because from biblical times this word meant *grain,* although to distinguish corn from other grains it was called *Indian corn.* In 1773, Linnaeus gave corn the botanical name of *Zea Mays; Zea* from the Greek, meaning *grain;* and *Mays (mais, maize)* from the aboriginal sound for *that which sustains life.*

The Corn-refining Industry

Until the middle of 1800s, starch was manufactured principally from potatoes and wheat. The corn-refining industry was started by Thomas Kingsford, in Oswego, New York; its first product was table starch. Kingsford's early efforts encountered much opposition and ridicule, but a huge industry has grown from his modest beginnings. Americans took an unquestioned lead in building this industry, and today it has spread to practically every country in the world. Total annual corn utilization by the U.S. wet milling process now amounts to about 250 million bushels, approximately 5% of U.S. corn consumption. In several states of the Midwest, corn purchases by corn refiners amount to 20% or more of the cash crop. Corn refining is essentially a two-part operation: (1) wet milling and (2) conversion. The milling process separates the kernel into its principal parts; the conversion transforms *starch,* the milling product, into syrups, sugars, and further derivatives.

The Corn Kernel

Corn refiners process the corn kernel and do not work with any other component of the corn plant, such as the stalk, leaves, husks, cobs, or tassels. In most instances, only one type of corn is used in wet milling—hybrid yellow dent corn. The objective of corn wet milling is to recover starch, which comprises about 70% of the dry substance of the corn kernel, although important refining by-products are also produced. No part of the kernel is wasted.

A kernel of corn has four principal parts. In order of their separation in wet milling, they are: (1) the germ; (2) the hull; (3) the gluten; and (4) the starch. The *germ,* containing most of the oil, is clearly visible to the eye. Also called the embryo, it extends from the inner end of the kernel, where the kernel joins the cob, well up into the body of the grain. The *hull* is the outer covering, or pericarp, of the grain; it is largely fibrous and is sometimes called the *bran. Gluten* and *starch,* less easily distinguished, are intermixed throughout the endosperm. Before processing, the gluten-rich endosperm is a hard, flinty, yellowish substance, somewhat translucent in appearance, just under the hull and mostly along the sides of the kernel. It contains most of the protein of the corn kernel. The starch-rich endosperm is whitish and softer, filling the broad upper area of the kernel and extending downward around the germ. The starch inside the corn kernel occurs in granules 10-30 microns in diameter and is composed of two major fractions which differ in chemical structure. One, the *amylose fraction,* has its molecules arranged in a linear or straight chain. The molecules of the other, the *amylopectin fraction,* are arranged in a branching chain. Commercially milled yellow dent corn starch has 26% amylose and 73% amylopectin starch.

Waxy corn (also known as waxy maize) contains starch that is virtually 100% amylopectin. This type of starch tends to promote stability in solution and resistence to gelling—useful properties in such foods as salad dressings and pie fillings. For some time, research has been conducted to develop corn of higher amylose content, and the ultimate goal is a type of corn with 100% amylose starch. The growing use of edible food films and wrappings offers a potentially great market for such a specialty corn.

Plant genetics may restructure the corn kernel in other ways as well. One new development is high-lysine corn, a corn altered by genetics, which contains higher amounts of the essential amino acid, lysine. Though some years away from commercial uses, high-lysine corn promises to unlock more of corn's protein for both animal and human nutrition. A long-standing goal has been the development of corn of higher oil content, and geneticists are also working on higher protein corns as well as a combination high-lysine high-protein corn.

Selection of Corn for Wet Processing

The problem of purchasing corn for the wet processor of corn is considerably different from the problem of the flour miller. The two principal guides in purchasing corn are the federal grain standards and personal inspection of the grain by the buyer. All colors can be used, but the great bulk of corn used by the wet corn processor is Yellow corn, because most corn entering cash grain channels is Yellow corn (White corn is generally at a premium over Yellow corn), and because Yellow corn produces gluten feed that causes the golden skin on iced broilers and more golden egg yolks, which consumers prefer. This is due to presence of the trace element xanthophyll in Yellow corn and its absence from White corn.

The factor of greatest importance to the wet corn processor is the amount of starch that can be recovered from the corn. Another factor is the condition of the germ, which determines the amount of oil that can be recovered. Although almost any grade of corn can be used in the wet milling process, the better grades result in a much greater recovery of products. The buyer for a wet corn processor knows the yields of products that may be obtained from the different grades of corn. Thus he has price differentials in his mind that he will and can pay for different qualities.

Careful attention is paid to the grade assigned to a particular lot. The buyer is interested in the percentage of damaged kernels, moisture content, foreign material, test weight per bushel, and the general appearance of the corn. In wet milling processing of corn, the grain is crushed and torn apart. All foreign material is separated from the kernel at this time, although the quantity and character of the foreign material must be taken into account in naming a price.

The two most important grading factors to the wet corn processor are moisture content and test weight. When the moisture content is high, it is less valuable to the buyer because he is obtaining less corn and more water. New corn may be of a high moisture content, and a particular lot may grade No. 5 because of moisture content, although all other factors might make it No. 1 or No. 2 grade. Most wet and dry millers prefer corn in the 17.5%-18.5% moisture range during the winter months. The test weight indicates the fullness of the kernel and the soundness of the corn. A high test weight is frequently indicative of a high-quality, full-bodied corn. By examining the corn and properly appraising it, the buyer is partially able to check the factors indicated in the grade.

The Wet Milling Process

The wet milling process begins when incoming corn is received at the plants in boxcars or hopper cars. Each car of corn is unloaded into a grated pit. From there the corn is elevated to temporary storage bins, then to scale hoppers for weighing and sampling. It next passes through the first clearing stage to separate unwanted substances such as pieces of cob, husks, sticks, sand, and stones. After cleaning, the shelled corn may be routed to storage silos as a reserve stock, or it may enter directly into the milling process. Corn that is cleaned and then stored is again cleaned mechanically (second corn cleaners) before processing begins. The first step in the process is the separation of the corn into its constituents—solubles, germ, hull, gluten, and starch.

A. Steeping

Shelled corn is *steeped* (soaked) from 36 to 48 hours in tanks called *steeps,* which hold 2,000 to 6,000 bushels. Warm water, treated with sulphur dioxide to retard undesirable fermentation and to aid in extracting soluble materials from the corn, is circulated through the steeps. The diluted sulphurous acid also softens the hull and loosens the starch-protein complex of the kernel so the starch and other components can be separated more easily. When the solubles content reaches a certain concentration, the steepwater is drawn off to vacuum evaporators, where it is concentrated to a syrupy liquid called a variety of names—*heavy steepwater, condensed fermented corn extractives,* or *steepwater concentrates.*

Because of its valuable store of protein and minerals, steepwater is important nutritionally. Much of the by-product is added to the hulls and gluten, which emerge later on in corn refining, to form livestock and poultry feeds. Since World War II, concentrated steepwater has become increasingly important as a nutrient in the microbiological production of antibiotic drugs, such as penicillin and Aureomycin, and B-complex vitamins.

B. Separating the Germ

After the steeping process is completed, the softening kernels are shredded in degerminating mills equipped with two close-set plates having projecting metal teeth that revolve in opposite directions. In this step, the corn is torn into coarse pieces and the soft, pliable germ is freed without being crushed.

From the degerminating mills the corn passes to flotation tanks called germ separators, or into centrifugal hydroclones, a newer process. The fine particles of starch and other components of the kernel turn the water into a milklike slurry, float to the surface, and are skimmed off at one end of the separators in a continuous stream. At the same time, the heavier starch and other components of the kernel are drawn from the bottom of the germ separators. The germ, together with some of the slurry that passes with it from the top of the germ separators, is carried through screens or reels which separate the starch and water, leaving the germs ready to be sent to the oil-processing section of the plant.

C. Oil Extraction and Refining

In the extraction of the oil, the germs are washed to remove adhering particles of starch, then squeezed to remove water, and dried in revolving cylindrical dryers. The germs are then treated under pressure and heat to aid in the removal of the oil. The germs thus prepared are run through expellers, where the oil is extracted by means of heavy mechanical pressure. The residue is ground for animal feed, either to be sold separately as oil meal (corn germ meal) or to be mixed with gluten, fiber, and solubles to make corn gluten feed. At this point the oil can be marketed as crude corn oil or can be further refined to produce edible oils by processes to be described later.

Some refiners use a combination of expellers and solvent extraction. In this process, the expellers are used to remove part of the oil and to reduce the bulk density of the germs, which come from the expellers in flake form. The final extraction step generally employs hexane, a hydrocarbon fraction of petroleum, in a countercurrent flow with the germ. The corn oil is soluble in the hexane and is thus extracted from the germ.

The solvent-oil solution from the extractor is filtered to remove pieces of flake, and then the solvent is separated from the oil in an evaporator and stripping column. The solvent boils off, after which it is condensed and collected for reuse. The germ flake coming from the extractor also contains a certain amount of solvent; this is recovered by a steaming or heating process before the flake, now stripped almost entirely of its oil content, goes to the feed department.

Oil from the solvent extractor is either filtered to remove suspended solid matter or run into large tanks where the solids settle to the bottom and are later removed. These solids, known collectively in the industry as *foots,* are used to manufacture soap and glycerin.

Crude corn oil can be used in soap making, paints, and varnishes, and in the manufacture of rubber substitutes. Most corn oil is not sold in crude form, however, because of its value for edible purposes in refined form. The breakdown of corn oil usage is as follows: salad and cooking oil (including some related food uses), 57%; margarine, 34%; shortening, 3%; soap stock and refining loss, 6%.

To produce refined corn oil, the crude oil is subjected to processing that involves; first, alkali treatment to neutralize free fatty-acid soap stocks; then, centrifugation to remove soap stock, leaching out the alkali residue from the previous process with activated carbon or some other absorbents, and blowing with live steam under high vacuum (known as *deodorizing*) to remove volatile odoriferous components; and, finally, chilling and filtering (known as *winterizing*) to remove stearin (fat solids) and waxes.

The soap stock may or may not be acidified to a liquid product. It may be used for soap, although its major use is now for tanning leather or as a softening agent in textile sizes. The acidified product is used as a fat source in animal feeds and is sold under the name of hydrolyzed vegetable oil.

D. Separating the Hulls

The slurry of hulls, gluten, and starch drawn from the germ separators passes through mills which thoroughly disintegrate the harder portions of the corn. For many years, buhrmills, of practically the same description as stone mills of ancient times, were used. Today, however, buhrstones have been replaced by stainless steel plates. The hulls are separated from the starch and gluten by means of reels or shakers covered with nylon bolting cloth. The extremely fine particles of starch and gluten are sifted through the cloth, leaving the hulls within the shakers to be discharged at one end. The hulls then pass on to be dried and become part of corn-gluten feed.

E. Separating the Gluten from the Starch

The wet milling process has at this point disposed of all parts of the corn kernel except the starch and gluten, which remain in suspension in the *slurry* after passing through the shakers. The next step is to separate the starch from the gluten by centrifugation. The contrifuges are like cream separators, with the lighter gluten particles flowing out at the center and the heavier starch emerging at the periphery.

Most of the gluten is used in corn-gluten meal, an ingredient of livestock feeds and particularly of poultry rations. Some of the gluten is added to the concentrated steepwater, corn-oil meal, and hulls to produce corn-gluten feed. Both corn-gluten meal and corn-gluten feed are long-established, high-protein ingredients of mixed feeds. Another use for gluten is to produce *zein,* a versatile, purified protein of corn used as a basic plastic material, as a carrier for textile colors, as a binder for abrasives, and as a laminating glue. Its consumer products include drugs and medicines, floor finishes, food coatings, industrial paints and lacquers, paper, photographic film, and printing inks.

After separation from the gluten, the starch, still in a slurry, is filtered and washed several times in order to remove any remaining trace of nonstarch materials. After the final filtering, the starch may be dried and packaged for market or given further treatment to produce specialty starches, dextrins, syrups, or sugars.

Wet Milling Products

Products derived from the wet milling of corn enter into livestock and poultry feeds, food, and nonfood industrial uses in hundreds of forms and applications. Research is constantly bringing to light new products and uses of corn, particularly the use of corn-derived products in nonfood applications.

A. Corn Starch Products and Dextrin

An almost endless variety of products is made from corn starch. Large quantities, of course, are prepared for human consumption. Much of this is milled and is sold in the form of powder. Edible starch is found in human food in many forms, including sauces and gravies, puddings and other types of desserts, gumdrops and other confections, and in the making of baking powder. Large quantities of starch are used in brewing beer and ale.

For other than edible purposes, special kinds of starch are produced by soaking the corn starch in water with an admixture of small amounts of acid (later completely neutralized), or with oxidizing substances. These modifications of starch find various uses in the textile industries, in the

Wet Milled Corn Products and Their Uses

	foods	industrial	livestock feed	brewing
corn starch	baking powder confections desserts gravies sauces	(dextrins) adhesives ceramics inks metal castings paints textiles	corn-gluten meal	ale beer
corn syrup	bakery products canned fruits confections food sauces frozen fruits frozen desserts ice cream jellies preserves table syrups	leather tanning paper pharmaceuticals textiles tobacco curing		
corn sugar	baby foods bakery goods canned fruits condensed milk frozen fruits	chemicals pharmaceuticals	mixed rations	
crude corn oil	bakery products condiments confections margarine salad dressings table oils	leather tanning paints soaps synthetic rubber textile sizings varnishes	corn-oil meal	
steepwater concentrates		pharmaceuticals		

manufacture of paper, in the preparation of adhesives, in fillers for cosmetics, in high explosives, as core binders in foundries, and for various laundry purposes. Raw starch is insoluble in water but, by physical or chemical treatment, it can be so altered that it will mix with water and form a solution. In this form, it is not necessary for the customer to cook the starch before the final end use.

Starch may be refined into dextrin, another commercially useful product. *Dextrin* is made from starch through a process of roasting dry starch, sometimes with the addition of a small amount of mineral acid. British gum, a type of dextrin, is made from dry starch by a roasting process without the addition of acid. Dextrins are varied to meet specific needs by changes in the length of time and temperature of roasting and in the amount of acid used. Dextrins, including British gum, are used to make adhesives of various kinds and are also used in the textile industry as sizing to strengthen the fiber and to add to the finish in cloth, carpets, and twine; for thickening the fixing agent for colors in textile printing; and even in the preparation of inks. Dextrins and starches are also used as core binders in the casting of iron, steel, brass, and aluminum, and in ceramics. A new class of dextrins made by partial enzyme hydrolysis, called *malto-dextrins,* have recently found versatile uses as water-soluble, nonsweet material in food products.

B. Corn Syrup

Corn syrup is manufactured from corn starch through a process of heating under pressure in large closed vessels called *converters.* In the process a small amount of hydrochloric acid is added; this is the same acid as that of the human stomach, but the concentrations used in producing corn syrup are much lower. The action in the converters is almost the same as that which takes place in the human stomach when starch is digested, with the starch being converted into sugar by hydrolysis.

Complete hydrolysis of starch yields *dextrose,* a simple sugar identical with the sugar of human blood. In the manufacture of corn syrup, however, the reaction is halted when only partially complete. As a result, only part of the starch is converted into dextrose; the remainder consists of maltose and higher saccharides. The proportions control the sweetness and physical properties of the different corn syrups.

Two methods of producing corn syrup are in common use: the hydrolysis process just described and the acid-enzyme process. In the latter, acid hydrolysis is followed by conversion with a purified diastatic enzyme, which results in a syrup of relatively higher dextrose and maltose content than syrup produced by the acid process alone. The initial acid-conversion step is the same for both processes. When conversion has reached the desired point, the hydrolyzate, as it is called, is neutralized by adding a small amount of sodium carbonate (soda ash) in the neutralizing tank. This procedure stops the conversion reaction.

Fatty substances that rise to the surface are either skimmed off or removed in fat-separating centrifuges, and suspended solid matter is removed by filtering. The filtrate then is evaporated to about 60% dry substance. After this initial evaporation, the hydrolyzate is passed through activated powdered carbon filters that effect further clarification and decolorization. This process removes colored organic substances, and the resulting syrup is as clear and colorless as water. Inorganic substances contained in the hydrolyzate are in *ionized* (electrically charged) form; these can be removed from the solution by ion-exchange systems.

After final filtration, part of the residual water is evaporated so that the syrup will have the desired density. Evaporation is carried out in large vacuum evaporators allowing the moisture to be removed at temperatures below levels that would result in thermal damage to the syrup. The syrup is next cooled and is then ready to be loaded into tank cars, tank trucks, or steel drums.

Where acid-enzyme or dual-conversion syrups are being made, the enzyme is added to the hydrolyzate after its initial evaporation to 60% dry substance. When the enzyme hydrolysis has progressed to the desired level, the enzyme is inactivated. Further refining and final evaporation follow as in the production of straight acid-conversion syrup.

The uses of corn syrup are many, one of the most important being the manufacture of confections. Corn syrup is especially suitable for this because it inhibits the crystallization of sucrose and other sugars. Corn syrup is made into a variety of table syrups through the addition of sucrose, molasses, sorghum, refiners' syrup, or honey, in order to give it a more distinct or pronounced flavor. Corn syrup is also employed in making jams, jellies, and preserves, and in canned and frozen fruits. In addition, it enters extensively into the manufacture of ice cream and other frozen desserts, bakery products, mincemeat, and food sauces. Nonfood applications of corn syrup include the tanning and finishing of leather, the manufacture of rayon, textile finishing, moisture conditioning of tobacco, the manufacture of glassine and parchment paper, drugs and medicines, and many other uses.

C. Corn Sugar

Refined corn syrup, or dextrose, is produced from starch by adjusting the conditions for hydrolysis so as to give a maximum dextrose content in the hydrolyzate. After purification and concentration, as in the manufacture of corn syrup, the highly refined dextrose liquor is seeded with dextrose crystals from a previous batch. Crystallization is then carried out under a carefully controlled rate of cooling in large crystallizers equipped with agitators.

After approximately 100 hours in the crystallizers, a *crystalline dextrose monohydrate* (one molecule of water held in the crystal with each molecule of dextrose) is formed. This is removed from the *magma* (mixture of crystals and mother liquor) in centrifuges, washed, dried, and packed in 100-pound paper bags. The mother liquor separated from the crystal by the centrifuges is reconverted, clarified, concentrated, and then passed through another crystallization cycle. As before, the dextrose crystals that form during this second cycle are removed from the magma by centrifuges. The dextrose obtained during this second cycle is redissolved in water, added to new starch hydrolyzate, and recrystallized to give pure dextrose monohydrate. The mother liquor from the second crystallization, dark brown in color, is called *hydrol* or *corn-sugar molasses* and is widely used in animal feeding and in several industrial processes.

Dextrose is used in the manufacture of confections, carbonated beverages, ice cream, various bakery products, baby foods, condensed milk, and fruit canning, as well as other food uses. In highly purified form, it is regularly administered intravenously after surgical operations, in cases of severe shock, and for other medical purposes. Dextrose also serves as a base for the manufacture of lactic acid, sorbitol, and mannitol, which are industrial chemicals of growing usefulness.

Anhydrous dextrose, which contains no water, may be made by redissolving and recrystallizing the dextrose under carefully controlled conditions. The resulting crystals are dried until they are practically free from moisture. This product is used for making chocolate coatings and various other foods requiring a water-free sugar.

15

other grain milling, processing

Dry Milling of Corn

Long before Europeans came to America, Indians were grinding corn for use as food. The methods employed by the American natives in preparing corn for consumption bear little similarity to modern milling methods, yet the basic idea of reducing and breaking down the corn by grinding is the same.

Unlike the wet processing corn industry, which is made up of only a few processors, the dry processing corn industry is comprised of hundreds of mills both large and small, scattered throughout the states where corn is produced. On an average, a little less than 3% of the annual corn crop is used in the dry process mills as compared with about 5% utilized by the wet process mills. The per capita consumption of dry processed corn has declined for many years as a result of slow changes in eating habits. The U.S. per capita use of corn meal in 1899 was 100.3 pounds; 70 years later it had dropped to 7.4 pounds.

A. Selection of Corn for Dry Processing

The dry milling processor of corn uses both White and Yellow corn. He is not able to use Mixed corn, however, because consumers demand the products to be either pure white or pure yellow. The dry processor is often compelled to pay premiums for White corn, owing to the small production of White corn. Farmers have preferred to raise the yellow hybrid varieties because of the high yields obtained. Today, nearly all corn produced in the U.S. is Yellow corn.

The purchasing problems of the dry milling processor of corn differ from those of the wet milling processor in the restrictions regarding grades of corn. In buying corn for the dry milling process, pure Yellow and pure White corn are specified. The corn must be plump, sound, cool, and sweet, and must contain only a small percentage of foreign material and broken kernels. The dry milling processor usually requires that the test weight of the corn be 52 pounds or over. Ordinarily the processor uses Nos. 1, 2, and 3 grades, but Nos. 4 and 5 grades can be used if they are these grades solely because of a high moisture content. Naturally dried corn is preferred to the artificially dried product, and the thick kernel is preferred to the thin. The dry milling processor attempts to avoid round kernels because they are more difficult to process, and grain that contains kernels damaged by mold or fermentation because of the possible effect on the quality of the finished product.

The process of dry corn milling is closely akin to the processes and operations of flour milling; the essential differences between the two is the difference in the objectives of the two processes. In flour milling, the reduction is carried to a point where as much raw material as possible is reduced to flour. In corn milling, the products of the early stages or reduction are worth more than the products of the final reduction.

In the dry milling processing of both Yellow and White corn, the general milling procedure is the same, although some products are not prepared from Yellow corn. The typical dry milling products derived from Yellow corn are granulated corn meal, bolted corn meal, brewers' grits, hominy grits, and flour; from White corn, pearl hominy, brewers' grits, hominy grits, granulated corn meal, bolted corn meal, and flour.

Corn that is to be dry milled is cleaned in much the same way as is wheat to be milled. After the corn is cleaned, the first step in processing is *tempering,* which is performed by treating the grain with hot water and steam. The purpose of tempering the corn is to soften the bran and germ so that they can be separated from the remainder of the kernel; the operation usually requires from two to six hours.

When the tempering is completed, the grain is carried to the *degerminator,* a machine that separates the germ and bran from the remainder of the kernel. The fine material is bolted out (sifted), and the remaining mixture of cracked corn, bran, and germ is fed into a separating machine, where the bran and germ are removed from the cracked, flinty portions. The bran is used in hominy feed. Corn oil is extracted from the germ, and the residue forms the by-product from which corn-oil meal is made.

After the bran and germ are removed, the flinty, coarse, granular material is further processed. First, this material is dried and cooled. The dried material, which is made up of pieces of kernels not uniform in size as a result of the original cracking, is graded according to size through the use of bolting and sifting machines. The largest pieces are used in the manufacture of pearl hominy, and the only additional process for this material is pearling. In *pearling,* the rough edges of the particles of the kernels are removed and the product is polished. Pearl hominy is used in the home as hominy, or is further processed in the manufacture of cornflakes through flavoring, cooking, rolling, and toasting.

The smaller pieces of corn extracted from the dried material are moved through a series of corrugated reduction rolls similar to those used in flour milling, but larger. After going through each corrugated roll, the material is graded according to size through the use of bolting and sifting machines. At each state of the reduction process, certain sizes are ready for packaging, while the others go on to further processing. The fine material resulting at each state is separated and used for animal feed.

The first product derived after the reduction of the material by the corrugated rolls is *brewers' grits,* used in the manufacture of beer. Brewers' grits are approximately one-fourth the size of pearl hominy. The next smaller grade of product derived through the reduction process is known as *hominy grits,* which are approximately one-fourth the size of brewers' grits. Hominy grits are packaged and sold for home consumption. *Granulated corn meal,* one of the best-known products of the dry processing corn industry, is the next product obtained. The next finer product is *bolted* or sifted *cream meal,* used for home consumption and certain industrial uses. The final product in the reduction of corn by the dry process is *corn flour,* which is used for home consumption, in industry, and for animal feed.

Oats Milling

Oats have been used for food for centuries in years when the wheat harvest failed. The development of the Central European civilization saw oatmeal introduced into the diet. In some areas with a cool, moist climate suitable to oats cultivation, oatmeal came to be an important element of food, and in Scotland oatmeal became probably the most important cereal. The milling of oats in the U.S. began during the middle of the 19th century, when a German immigrant, Ferdinand Schumacher, started a mill at Akron, Ohio. Since that time, products of oats milling have attained an important place in the American diet.

Physically and chemically the oat grain is in many respects similar to wheat, barley, and rye. Like other grains, the coating of the *groat,* the oat grain, is known as the *bran* or *hull,* and like other grains the oat grain also contains a germ or seed. Unlike the bran coat of the wheat kernel, the oat hull is not firmly attached to the groat and can be removed with comparative ease. The oat grain contains one-third more protein than wheat and nearly four times as much fat, but less starch.

The edible products of processed oats are rolled oats, steel-cut oatmeal, and ground oatmeal. For many years after the introduction of oat milling in the U.S., oat hulls, a by-product of the milling process, were sold for any price they would bring or were given away or destroyed. With the development of the mixed-feeds industry, however, a market was developed for these hulls.

An additional outlet for oat hulls has been developed in recent years with the discovery that they can be used in the manufacture of *furfural,* a liquid with many existent and potential uses. The first large-scale use of furfural was in the manufacture of synthetic resins, such as Bakelite, and this is still one of the major outlets, with a promise of becoming still larger in the future. More recently, large quantities of furfural have been used for purifying dark-colored wood resin and for refining lubricating oils. Scores of other uses for furfural could be cited, including its use as a fungicide and a solvent for lacquers and varnishes, in the textile printing field, and for abrasive wheels. Successful experiments have used furfural in the construction of bituminous- and asphaltic-type roads.

A. Selection of Oats for Milling

In the selection of oats for milling purposes, White oats are preferred. The oats miller cannot rely solely on federal grades but must also inspect the grain carefully. Size and uniformity of kernel and absence of admixture of other grain, particularly barley, are important considerations. Samples of the grain are judged on the basis of estimated outturn of finished products and estimated percentage of materials such as small kernels, oat seeds, chaff, and other seeds. The test weight per bushel is not always a satisfactory indication of milling quality, although a 32-pound per bushel test weight or higher is preferable to the lighter test oats which prevail in some crop years. The oats miller is also interested in obtaining grain that does not run much over 14.5% moisture content.

B. The Oats-milling Process

The cleaning and preparation of oats for milling is an intricate process requiring about 85 steps, including separating by aspirators, grading or screening as to diameter and length, and blending. Chaff, light (or pin) oats, double oats, hulled oats, and other grain and seeds are separated by special machines, with the result that only perhaps 50% to 75% of the original weight of the grain remains for the milling process.

The oats thus separated as suitable for processing are next carried to the drying house, where they are dried in a series of pans (usually about 12) grouped in vertical columns. The oats travel from the top pan to the pan beneath, until they reach the bottom pan. While moving from top to bottom the oats are stirred in each pan by a large mechanical sweeper. As a result of the drying process, the kernels lose between 6% and 7% of their weight through loss of moisture, and, at the same time, they acquire a flavor because of the toasting and drying. After drying, the oats are cooled and then placed in tempering bins for a period of from 12 to 24 hours.

Next, the oats are graded into three classes—large, slim, and stub—which are milled separately. The oats are hulled through the use of two milling stones, one of which revolves and the other of which is stationary. Following the hulling process, the oats are carried to cleaning machinery consisting of reels that bolt out the dust, and aspirators that remove the hulls. Kernels that have not become separated into hulls and groats at this stage are returned to rehulling stones.

Groats destined for steel-cut oatmeal are cut by special steel rolls into granular pieces; they are then ready for packaging. The groats intended for rolled-oats production are passed into a steam chamber and partially cooked by live steam, after which they go through rollers where they are formed into flakes. The rolled-oats flakes are then cooled in a current of air to about 100° to 120° F and are immediately weighed and packed by automatic equipment.

Mixed Feed Manufacturing

The U.S. feed-manufacturing industry, which originated in the 1890s, has grown from a handful of grain and by-product mixers to the 13th largest manufacturing industry in the U.S.—the largest industry in the world devoted exclusively to supplying goods and services to agriculture.

About 6,000 manufacturing plants and thousands of dealers, employing over 100,000 people and representing a total investment of nearly $2 billion, make up this industry, whose annual gross sales are nearly $4 billion. The annual production of this industry was almost 30 million tons of livestock and poultry feed in 1950; by 1970 this had increased to 55 million tons. The rapid growth of this industry has been due to the ever-greater need and desire for animal products and the search for greater efficiency in their production. An additional factor is the trend to larger farming units. In the case of livestock and poultry units, great efficiency has been achieved, in part due to scientific feeding. In the 1890s, stock roamed loose and subsisted on what could be found growing in the pasture, plus table scraps and some grain. Today's stock is confined and fed relatively complex diets—manufactured from 20 to 30 different ingredients.

A. Ingredients

The ingredients used in producing feeds are grains, such as wheat, corn, oats, barley, and milo; oil seed meals, such as soybean, cottonseed, and linseed meals; meat scraps and tankage; fish meal and solubles; grain by-products, such as middlings, bran, gluten feed, and meal; brewers' and distillers' grains and elevator screenings; alfalfa meal; molasses; minerals and trace minerals; vitamins; antibiotics; drugs; and additives.

As much as 85% of some manufactured feeds is made up of grain and/or grain by-products. While grain is the backbone of manufactured feeds, it must be fortified or reinforced by other ingredients to produce a balanced diet for livestock and poultry. In formulating feeds, the nutritionist first establishes the levels desired of each nutrient factor for the class of livestock to be fed. He will use as his source of information the research work done at agricultural colleges and by industrial and governmental research. Having established the levels of nutrients wanted, he will compare these with the levels known to be contained in grain and other ingredients. A combination of the ingredients is then determined that will meet the nutritional requirements economically, be palatable to the animal, capable of being processed, and acceptable to the customer.

In brief, a combination of ingredients is sought that will result in optimum animal production coupled with minimum feed cost. It is each feed manufacturer's goal to do a more efficient job of producing a superior product than his competitors and thus to create a demand for his feeds.

B. Types of Feeds

A feed is classified by the livestock for which it is designed: (1) poultry, (2) hog, (3) cattle and dairy, (4) sheep, and (5) specialty and small animal.

This classification may be further divided into the stage of development or purpose for which it is fed, such as starter feed, growing feed, finishing feed, breeder feed, and production feed.

All feeds fall into one of two general categories: complete feeds or concentrates. In brief, *complete feeds* are a balanced ration in themselves, and nothing else, except water, need be fed to the animal. These feeds contain high levels of grain and grain by-products. *Concentrates* are feeds in which the grain, and possibly one or two other ingredients, have not been included; the missing items are usually added at the farm before feeding. Most concentrates contain high levels of high-protein soybean, linseed, and cottonseed meals, and serve as supplements to grain grown on the farm.

Slightly over half of all feed manufactured in the U.S. is poultry feed. Swine feeds amount to less than one-fifth of all production, and cattle feeds make up a little over a quarter. While most poultry feed is complete feed, swine feeds are over two-thirds concentrates. Cattle feeds are almost completely concentrates.

Feeds are also distinguished by their physical form. They can be (1) meal—all ingredients ground or processed into small particle size, (2) pellets—meal which is formed into pellets, or (3) crumbles—reground or cracked pellets. In some cases, crimped or cracked grains and other relatively coarse ingredients are used to make feeds of coarse texture that are desired by the dairy industry.

C. Mixing

Most ingredients arrive at a feed mill in a form ready to be mixed. Grain is the exception; it is easier to handle and keeps better in whole form. Grain is handled in whole form until ready for mixing, at which time it is ground into meal.

The mixing of the various ingredients in manufactured feeds is no longer the muscle and scoop-shovel task of past years but is the job of skilled men controlling expensive, specifically designed equipment. Today the complete operation—unloading inbound ingredients, transporting and storing them within the mill, mixing them within close tolerances in various ratios to one another, packaging the mixed feed, transporting the feed to the loading dock, and loading the feed in railcars or trucks—is done in many modern mills at a fraction of a man-hour per ton of feed produced. Practically the whole process is mechanized.

Probably no two feed mills in existence are exactly alike in design or equipment. Just as many different formulations of feed will give comparable results when fed, there are many ways of manufacturing feeds. In general, however, feed mills use the batch system, the continuous-mixing system, or variations or combinations of these two systems. In the *batch system,* all ingredients are weighed and dumped into a mixing system, and ingredients are volume-metered into a mixing screw conveyor. High production volume on one type of feed favors this system.

In most larger plants, the majority of ingredients arrive in bulk by rail or truck. Smaller plants utilize more truck transportation and receive more ingredients in sacked form. Outbound feed moves in both sacked and bulk forms via rail and truck.

Barley Malting

Malt is a grain whose starch content has been converted to a form of sugar through a process of sprouting or growing. It is used in the brewing industry as well as in the distilling industry.

The barley malting industry in the U.S. is confined to a relatively few firms. Most brewers and distillers buy their malt from these maltsters, although a few brewing firms make their own. The malting industry uses approximately 30% of an average annual U.S. barley crop. Roughly speaking, one bushel of barley (48 pounds) will produce 1.15 bushels of malt (34 pounds), which, together with such additions as corn, rice, and hops, will produce approximately one barrel of beer.

A. Selection of Barley for Malting

The malting industry is exacting in the type of barley that may be used, and only certain types of barley are acceptable for its purpose. Because of these rigid requirements, barley suitable for malting usually commands a substantial premium over barley used for other purposes. Malsters ordinarily use what are known as 6-row varieties of barley for malt to be used in the preparation of the lighter types of beers, while they use 2-row varieties for the heavier beers and ales.

The characteristics of barley used for malting in the U.S. vary considerably from those of the barley used in the United Kingdom and Germany. The variance in the character of the raw material arises because German and United Kingdom brewers are obliged to supply a demand for dark, heavy beers, while the U.S. custom demands a light, pale beer. Since Americans prefer the lighter beers, 6-row varieties are used to a great extent by domestic maltsters.

Among the requirements of barley for malting are soundness, mellowness, sufficient test weight per bushel, uniformity and proper size of kernel, dryness, cleanliness, and color. Barley suitable for malting must be cool and sweet. Unsound barley will not germinate properly, and thus the percentage of sound kernels is also of importance. Barley is mellow when its starch is white and chalky, although the starch of some types of barley is colored, usually blue or gray, and hard in texture. The buyer is aided in determining soundness and mellowness through a pearling test. In pearling barley, the entire outer husk is removed in a scouring machine to disclose more readily the percentage of mellow and sound kernels.

Test weight is a factor in determining whether the kernels are well filled out and fully developed. Not only must malting barley kernels be well filled out, but they must be relatively uniform in size because kernels of varying size will not malt uniformly. Since the maltsters may find it necessary to store barley for several months, barley with excessive moisture is not desirable because of the risk of deterioration in storage. A final requirement is that barley used for malting must be virtually free of foreign material.

Variations in the climate, soil, and fertility affect the quality of barley. Seed deterioration, improper rotation of crops, shallow plowing, irregular seeding, and premature cutting and harvesting will affect the physical and chemical constituents of barley. Because of the numerous factors that may influence its quality, the quantity of barley available for malting varies considerably from year to year, and in some years it is difficult for the maltster to secure ample supplies of malting barley suitable for his requirements.

As in the case of wheat, the section of the country from which barley comes is an important consideration in buying. Early in the year, maltsters determine by tests where the best barley is available, and in purchasing supplies they attempt to get barley from these areas.

Malsters also rely on tests of grain to determine suitability for malting. A germination or growing test, which indicates if the lot has satisfactory germinating qualities, is made for each lot of barley. In addition, a so-called clean-out test is made, in which barley is sized into fancy, choice, and standard kernels. The remainder, which does not fall into these three classes, is known as the *clean-out.* These tests are made after the barley is purchased, and if they show that it is not satisfactory for malting, the barley is then sold for feed.

B. The Barley-malting Process

As in other grain processing, the first step in malting is that of cleaning the grain; this is followed by sizing so that uniform grains may be processed together. With the barley cleaned and sized, it is ready for the steeping process. The purpose of steeping is to moisten the grain sufficiently so that it will germinate. Water approximately 55°F is used in steeping, and the length of time required for steeping may vary from 30 to 72 hours, depending on the temperature and character of the steepwater, the character of the barley, and the size of the steeping tank. Correctly steeped barley contains between 46% and 50% water.

After being steeped, the barley is taken to the germinating or growing room. The floors of this room are made of perforated sheet metal, with the perforations not large enough for the grain to pass through and yet large enough to permit air to circulate among the grain. To prevent the sprouting grain from drying out, the humidity of the room is maintained at 96% to 98% during the growing period. The temperature in the room ranges between 50°F and 60°F. The temperature of the growing grain is above the room temperature, usually ranging from 60°F to 80°F, depending upon the character of the barley being sprouted and the type of malt desired.

When the barley begins to germinate, a stem starts growing just as if the barley were planted in the ground. By regulating the temperature and humidity of the room and of the growing grain, the maltster is able to control the rate of growth. Approximately six days are required for germinating most American barleys, and the germination is considered completed when the stem has grown from three-quarters to the full length of the grain. In the process of germination, the starch is converted to sugar.

During the later stages of the growing process, the drying or withering process begins. In *withering,* cool dry air is passed through the grain continuously, removing excess moisture without heat and at the same time modifying the starch body. *Drying* is accomplished when the sprouted grain is carried to the kiln, which is similar to the growing floor in that the floors are made of perforated sheet metal. In the kiln, hot air is drawn through the perforated floor and passes through the grain. Drying requires about 48 hours; and when it has been completed the finished malt contains between 3% and 4% moisture. By the time the malt has been sufficiently dried, the sprouts have become very brittle and easily break away from the kernel. At this point the sprouts are removed from the malt by screening, and the malt is placed in the storage bins for aging.

Distilling

The art of distilling has been known for many centuries, and some authorities claim that spirits were produced as long ago as 2,000 years before the Christian era. The first actual records of distillation date back to Arabian chemists, and the word *alcohol* is derived from the Arabic *alhuhl,* which is a powder used for painting the eyelids and obtained by sublimation, a process similar to distilling.

The technique of distilling has undergone improvement from the time of its discovery, with marked progress since the beginning of this century, although the basic procedure remains the same. Essentially the process consists of the conversion of starch into fermentable sugar, fermentation, and, finally, *distillation*—the process of separating the alcohol from the fermented mash.

A. Ingredients and Products

The principal grains used in the U.S. in the preparation of distilled liquors are corn, rye, and barley malt, although oats and wheat are also used. In addition to grain, another important ingredient in distilling liquors is yeast. The distiller is able to use most grades of rye and corn in his process, although some refuse to use any except the No. 1 and No. 2 grades. Distillers feel that the high quality of their product can be maintained only through the use of the better grades, and their rye and corn needs are filled chiefly on the basis of grade. The type of malt required by distillers differs from the type used by brewers. While brewers desire a malt that will produce beer that is palatable and tasty, as well as having good color and aroma, the principal requirement for malt used in distilling is that it be high in diastatic power, i.e., the capacity for converting starches to sugar.

The principal products of U.S. distillation are bourbon, rye, and corn whiskey. Corn and barley are the principal elements in bourbon whiskey. Ordinarily some rye is also used, but, in order to be classed as bourbon, at least 51% of the grain used must be corn. The same ingredients in different proportions are usually used in the preparation of corn whiskey. In producing rye whiskey, the principal ingredients are rye and rye malt or barley malt. Rye whiskey is also made with a portion of corn, but at least 51% of the grain must be rye to be classed as rye whiskey. Another important product of distillation is high-proof alcohol, or neutral spirits, which is used in the preparation of gin, in the manufacture of cordials, in blending, and in certain technical uses.

The Distilling Process

In distilling, as in all other grain processing, the first step is the cleaning of the grain. Next the grain is ground into meal, which is then conveyed to a bin where it is held until needed for processing. From the storage bins, the meal is placed in overhead scales where it is weighed by government inspectors. The meal is then dropped into the cookers and water is added at the rate of 18 gallons of water to one bushel of meal. The purpose of cooking is to rupture the starch cells and thus to facilitate the conversion of starch into sugar. When the mash has been cooked until the starch cells are satisfactorily broken down, cold water and barley malt are added. Malt has the capacity to convert the starch to sugar because of certain biological catalysts known as enzymes. The principal enzymes important to the distillers are diastase and protease. The *diastase* brings about the conversion of the cooked starch into sugar, and the *protease* breaks the proteins into simpler products that are used as yeast food in the fermentation process.

From the cookers the hot mash is pumped through a series of double-jacketed pipes, the outer pipe containing cold water running countercurrent to the hot mash being pumped through the inner pipe.

By the time the mash reaches the end of the pipe, it is cool. At that point, it enters the fermenting vats where yeast is added to bring about the fermentation process. The fermentation period lasts from 72 to 96 hours, and during this period the rising temperature developed in the mash is controlled by a cooling device. If the heat that develops during the fermentation process were allowed to rise unchecked, the quality and flavor of the liquor would be spoiled. In the distilling industry, fermented mash is referred to as *beer.*

After fermentation is completed, the beer, which now contains 10% alcohol, is pumped into a large tank where it is held until it is ready for distilling. From this reservoir tank, the beer is carried to the *still,* an apparatus for distilling the alcohol from the beer. There are different types of stills used in the distilling industry, but the most modern is the continuous type. Whatever the type of still used, the basic process is the same: Through the heating of the mash, the alcohol content is first vaporized and then condensed. The spent mash is pumped to large tubs where, through the use of screens and presses, a large part of the moisture is removed. By means of driers, the spent mash is completely dried and is then ready for use as livestock feed. Approximately 16 pounds of solid matter are recovered from each bushel of grain originally entering the mash.

After distilling is completed, the whiskey goes to a tank in the cistern room, where it is stored until it is placed in barrels. When the whiskey is put in barrels, government inspectors supervise the operation and check the weights. The filled barrels are placed in warehouses operating under government bond, where the whiskey is aged and matured. The purpose of storing whiskey is to modify its flavor, bouquet, and color. Whiskey is stored in charred, white-oak barrels. Over a period of time, certain qualities of the charred wood change the whiskey from a colorless liquid to the well-known amber color, and the carmelized gums and resins in the charred wood give flavor to the product.

The use of the charred oak barrel was discovered by chance shortly before the Civil War, when whiskey was made in the kitchen, and barrels were accumulated in the storage rooms connected with the house. Crude methods were used to clean the barrels and in some instances the interior of the barrels were burned out to reduce unpleasant odors and flavors from their previous contents. It was found that by using the charred oak barrels, whiskey matured and became more agreeable. Thereafter this process was recognized as the American standard of aging whiskey.

There is some difference of opinions as to the length of time that whiskey will continue to improve as a result of the aging process. The lighter whiskeys are believed by many persons to have reached their maximum qualities at the end of four years of aging; some of the heavier whiskeys are aged for a considerably longer period of time.

In a general way, the process of distilling high-proof alcohol is similar to that of whiskey. Gin is often made by mixing alcohol with flavoring extracts, such as juniper berries, orange peel, lemon peel, angelica root, orris root, and salt, but it is also made by a redistillation process after the flavoring extracts have been added. Gin differs from whiskey in that it does not require aging and is a finished product as it flows from the still.

While many imported vodkas are not distilled from grains but from fermented potatoes, most U.S. vodkas are distilled from either rye or wheat mash. Unlike gin which must have a distinct flavor, vodka must be essentially flavorless and this is accomplished by further processing of distilled neutral spirits.

In the manufacture of cordials and liqueurs, flavorings, such as mint and cocoa, and fruit flavors are blended with distilled neutral spirits.

C. Federal Supervision

An important factor in the distilling industry is the close supervision exercised by federal authorities. At the time the materials go into process, government inspectors supervise the weighing of the ingredients to determine whether the correct proportion of the different materials is used. When the whiskey is placed in barrels after the distillation process, another government inspector checks the weight. During the aging process the barrels of whiskey are stored in government-bonded warehouses; and before the whiskey or alcohol is bottled, the distiller must pay the federal tax through the purchase of revenue stamps which are affixed to each bottle.

16 soybean processing

Though the soybean has had a long history, agriculturalists showed very little interest in the crop until about 150 years ago. In 1904, the USDA published its first pamphlet on soybeans, but real interest in the crop developed only after W.J. Morse of the USDA sent many varieties and strains of soybeans back to the U.S. from the Orient in the late 1920s. These varieties were adapted to the U.S., and they soon flourished. By 1941, more acres were used to produce soybeans for seed than for hay, pasture, and other uses.

The soybean industry has almost single-handedly changed the U.S. from a net importer of fats and oils prior to World War II to its present position as a net exporter. The growth of the soybean industry has been an important factor in providing Americans with beef, pork, and poultry. In addition, the growth of the industry has benefited cash grain farmers with a soybean crop that was valued at $6.9 billion in 1975.

Soybeans have become the nation's leading export during the last 15 years. The total annual export value of soybeans during the 1975 crop year was placed at $2.488 billion. In the early 1970s, a number of factors produced increased demand for high-protein animal feeds, notably a tremendous shortage of feed grains in Eastern European countries, at a time when these nations were concentrating on increasing meat production. This trend to heavier production of meats is thought likely to continue and should mean continuing strong export demand for both soybeans and the processed meal.

Soybean Production

While environmental factors largely determine place of production, quantity of production is determined by the acreage planted and the yield. Planted acreage is determined by such factors as corn vs. soybean prices, government loan and support-price activities, and government land-retirement policies. Yield is determined by plant varieties and management practices. The rise in average per-acre yield from 11 bushels in 1924 to a record 28.4 in 1975 reflects the success of both agricultural scientists and growers and is a major contribution to the growth of the soybean-processing industry. The production increase from 555 million bushels in 1960 to more than 1.470 billion bushels in 1975 is, however, largely attributable to increased acreage rather than higher yields.

Demand for soybeans and soybean products has absorbed the production from over 50 million acres formerly planted to feed grains and other crops. Each year, the USDA attempts to adjust soybean production in terms of projected demand. One of the major determining factors behind acreage is the acreage-diversion policies of the USDA. These may take the form of special payments to promote or discourage planting of soybeans in relation to other crops such as corn or cotton.

The loan program applies to soybean production as well as to that of other major crops. However, demand for soybeans, soybean meal, and soybean oil in recent years has kept soybean market prices well above loan-rate levels, and there has been only small use made of the loan program. As a result, government-owned stocks of soybeans have been minimal. The government has acquired some soybeans in years of large production, but the program has been self-liquidating. In fact, during the last 10 years, use-has exceeded production, and carry-over has, at times, dropped to less than one month's requirements.

Processed Products

Soybeans have little utility as soybeans; their primary economic value is in the form of meal and oil. Soybean meal is produced either as a high-protein feed ingredient or for further processing into soy grits, flour, or isolated protein for use in human foods. Crude soybean oil is further refined and used in a variety of salad and cooking oils, margarines, prepared foods, and industrial chemicals. The following diagram indicates the variety of soybean uses and products.

	foods	feed	industrial	seed
whole soybeans	baked soybeans dietetic foods full-fat soy flour roasted soybeans soy butter soy cereal	puffed soybeans roasted soybeans steamed soybeans		hybridization planting
soybean oil	(refined soybean oil) cooking oil margarine mayonnaise prepared foods salad oil spreads (soybean lecithin) coatings emulsifiers nutrients		(refined soybean oil) adhesives disinfectants inks plastics (soybean lecithin) antifoam agents antiknock additives dispensing agents (sterols) (fatty acids) (glycerols)	
soybean meal	(basic meal) baby foods bakery beverages cereals confections dietetics meat analogs meat products soups (isolated proteins) (soy flour, grits)	(basic meal) millfeeds livestock feeds pet foods poultry feeds (soybean millfeeds)	(basic meal) fertilizers fillers (soy flour, grits) adhesives coatings	

Processing Methods

Approximately 120 plants in the U.S., primarily in the Midwest, process soybeans into oil and meal. Originally, soybeans were crushed by a hydraulic process similar to that used by linseed and cottonseed crushers, but this method did not perform well on oil recovery. It was replaced by the *expeller,* or screw, process, which removed between 90% and 95% of the oil. In the 1950s, the expeller process gave way to the solvent-extraction method.

Solvent-extraction plants now "crush" at least 95% of the soybeans processed in the U.S. In this method, the soybeans are cracked and flaked in special machines and conveyed through towers or tanks where they are leached with a chemical solvent. This hexane solvent dissolves 99% of the oil out of the flake and produces a mixture of oil and solvent, which is then distilled to recover the solvent for reuse, leaving pure crude soybean oil. The soybean flake, now minus most of the oil, is desolventized, toasted, and ground, producing soybean meal containing 44% protein or 49% protein.

Soybean Meal

Although 44% meal has been considered the standard meal product, a substantial demand has developed in recent years for a 49%-protein soybean meal, and today roughly half of U.S. soybean-meal production is of the 49%-protein product. To produce it, processors dry, clean, and dehull the soybean prior to flaking. The prime use for the 49% meal is in the more critical types of feeds, such as starter feeds and high-performance feeds. The hulls are a low-protein, high-fiber product and are used in bulky feeds or as carriers for molasses and similar materials.

Approximately 98% of the meal produced is used for poultry and animal feed; the remainder is used for human food or for industrial purposes. American feed manufacturers use soybean meal as a prime ingredient in mixed rations for poultry, hogs, and cattle to provide basic protein and amino-acid requirements for a balanced diet. Meal usage by types of animals is approximately as follows: poultry, 62%; hogs, 25%; cattle, 10%; sheep, other livestock, and pet foods, 3%. Processors watch developments in livestock and poultry prices to anticipate prospective demand for soybean meal.

Soybean Oil

In the refining process, crude oil can be degummed, refined, bleached, deodorized, or hydrogenated, depending on the use to be made of the oil. Degumming is accomplished by centrifugally separating the gums that have been precipitated from the oil by water or steam. The removed gums are then treated, and sometimes bleached, for various grades of lecithin, a product that has a widening field of uses in foods and industry. Oil is refined by neutralizing the free fatty acid with caustic acid and is centrifuged to remove the resultant soap stock. In some cases, degumming and refining are combined in one process. Oil is bleached by treatment with absorbent clay material, deodorized by subjecting it to steam distillation at high temperatures under vacuum, and hydrogenated by combining the oil with hydrogen gas in the presence of a catalyst.

Approximately 92% to 93% of soybean oil is used in foods, while the balance goes into industrial uses. The food industry uses soybean oil in mayonnaise, salad dressings, margarine, and vegetable shortenings. Soybean oil is used industrially in products such as soap, putty, inks, paints, and lubricants.

Soy Flour, Grits, and Isolated Protein

After the oil is removed in processing, the resulting product, which is called the *extracted flake,* can be ground and toasted to make meal. If the flake is dehulled, additional processing can produce soy flour, grits, and isolated protein. Coarse grinding of the dehulled extracted flakes produces grits, while a very fine grinding produces soy flour. These products have many food uses (e.g., baking, meat products, and cereals), and are used industrially for adhesives and as nutrients for antibiotics. Generally, the protein in flour and grits runs 50% to 53%. Isolated protein is obtained from the dehulled extracted flake by a chemical process that removes essentially all of the carbohydrate and fiber, concentrating or isolating the protein to 95% dry basis. This product has many food and industrial uses, including increasing protein content, reducing shrinkage, and providing structure and form to processed foods.

Gross Processing Margin

A critical factor in the rate of domestic processing of soybeans, which is the major determining factor affecting soybean supply/demand considerations and price, is the profitability of processing. The profitability is measured through the use of a formula called the *Gross Processing Margin* (GPM). This formula measures the cost of soybeans in relation to the sales return on processed soybean oil and meal. The GPM should be adequate to return to the processor his manufacturing costs, plus a profit.

The Gross Processing Margin translates the oil price per hundredweight and the soybean meal price per ton to the price for the equivalent yields of oil and meal from a bushel of soybeans. Though yields for both oil and meal will vary with soybean quality and processing efficiency from plant to plant, there is general agreement that a 60-pound bushel of soybeans will normally yield 47 to 48 pounds of soybean meal and 10 to 11 pounds of soybean oil, with a processing loss of nearly 2 pounds per 60-pound bushel.

In calculation of futures or *board crushing margins,* it is customary to use standard conversion factors that differ somewhat from usual physical product yields per bushel of soybeans. For the conversion example here, we will use 11 pounds of oil and 48 pounds of meal. Consequently, oil prices are converted to bushel equivalent by multiplying the oil price by 11, and the meal price is converted to a bushel equivalent by multiplying the meal price by .0235 (to save dividing the price by 2,000 and then multiplying by 48). The oil value and the meal value are added to obtain the product value or yield. The soybean cost is then subtracted to obtain the Gross Processing Margin.

Example:

Meal price	$126.00 per ton
Oil price	$0.24 per pound
Soybean price	$5.36 per bushel

$126 × .0235	=	$2.96	=	Meal value
$.24 × 11	=	+ 2.64	=	Oil value
		$ 5.60	=	Product value
	=	- 5.36	=	Soybean cost
		$ .24	=	Gross Processing Margin

Orders for execution of futures transactions are customarily based on the conversions shown above. Calculations of cash crushing margins by individual processors, however, will normally be based on actual product outturns expected.

Trading Techniques

There is frequently a seasonal pattern in processing margins, reflecting a favorable margin or conversion ratio during the early months of the crop year, following harvest, when soybeans are normally in heavy supply and at their lowest price levels. Nearly half of the soybean crop is sold in the first three months of the crop year (October, November, and December). Since oil and meal sales are seldom equal to soybean purchases during this quarter, however, some kind of price protection is needed by the processor. Soybean, oil, and meal prices, therefore, are analyzed as a whole by the processor, making the hedging program complex.

The existence of futures markets in soybeans and also in soybean meal and crude soybean oil permits the soybean processor to hedge against rising soybean prices and also against the risk of falling prices for processed soybean meal and oil. As in the grains, hedging against the risk of rising prices for soybeans is accomplished by purchases, long hedges, in soybean futures. Conversely, the hedge against the risk of falling prices for the finished products is accomplished through sales, short hedges, in soybean meal and soybean oil futures. It is most common for soybean processors to simultaneously hedge beans, meal, and oil by *putting on the crush*—a method used in periods of favorable bean and product cost/price margins—or *reversing the crush* when margins are not favorable. Though these three-way crushing and reverse-crushing methods are most commonly used, there are situations in which a soybean processor may not hedge in all three futures markets simultaneously.

Certain variations of this procedure take place when cash oil and/or meal are sold for deferred shipment. Various other techniques are also used, such as hedging cash soybeans in oil and/or meal futures.

A. Putting on the Crush

Futures markets for soybeans, oil, and meal permit soybean processors to put on crush or reverse crush. At times, the GPM that exists between the futures markets for soybeans, oil, and meal appears attractive. The processor might then use a hedging technique known as *putting on the crush.* For example, in the month of August, the GPM existing in January soybeans vs. January oil and January meal might be 27 cents. Processor X, after careful analysis, reasons that there is a strong likelihood that the GPM will narrow or worsen between August and the following January. He therefore would be inclined to put on crush, i.e., buy January soybeans and sell January oil and meal. By this action, Processor X attempts to establish his gross margin for this particular position. When cash soybeans are bought against the January crush, the long January soybean futures are sold out; as cash oil and meal are sold for January shipment, the short positions in oil and meal futures are covered. Assuming that Processor X was correct and the GPM did deteriorate from 27 cents, Processor X would be assured of 27 cents on the amount of crush he had on in the futures market.

B. Reverse Crush

Occasionally the cash market will be confronted with high soybean prices, reflecting tight supplies, and a processing margin that is unfavorable or negative. In this situation, the sales return on oil and meal in relation to the cost of soybeans is insufficient to allow for profitable processing. Rarely do plants shut down completely; instead, production slows down and this, together with reversal of hedging operations on the futures market, tends to correct the unfavorable cost/price relationship. The decline in soybean processing eases the drain on tight supplies and tends to curb rising soybean prices. At the same time, reduced processing will tend to lead to tighter supplies of oil and meal, increasing their price levels. The cost/price relationship tends to reverse, and favorable processing margins are reestablished.

In times of poor GPM's, the processor will use a futures trading technique known as *reverse crush*. In this procedure, the processor sells soybean futures and buys oil and meal futures. When the negative margin returns to a plus, the reverse crush will be closed out in the futures market, usually at a profit.

bibliography

Useful Sources of Information

Aldrich, Samuel R., Walter O. Scott, and Earl R. Leng. Modern Corn Production. 2nd ed. Champaign, IL, A&L Publications, 1975.

Inglett, George E., ed. Corn: Culture, Processing, Products. Westport, CT, Avi, 1970.

Inglett, George E., ed. Symposium: Seed Proteins. Westport, CT, Avi, 1972.

Inglett, George E., Wheat: Production and Utilization. Westport, CT, Avi, 1974.

Mortenson, W.P., and R.A. Luening. The Farm Management Handbook. 5th ed. Danville, IL, Interstate Printers and Publishers, 1972.

Sinha, R.N., and W.E. Muir. Grain Storage: Part of a System, Westport, CT, Avi, 1973.

Scott, Walter O., and Samuel R. Aldrich. Modern Soybean Production, Champaign, IL, S&A Publications, 1970.

Pamphlets

Wheat Flour Institute. From Flour to Bread: the Staff of Life. Chicago, Wheat Flour Institute, 1971.

Wheat Flour Institute, From Wheat to Flour: the Story of Man in a Grain of Wheat. Washington, D.C., Wheat Flour Institute, 1976.

Corn Refiners Association. The Corn Refining Industry. Washington, DC, Corn Refiners Association, 1967.

U.S. Government Publications

The USDA offers a broad range of publications on topics related to the economics of grains. Most of these are available without charge. For further information about their availability, you may write the appropriate department, listed below, in care of the U.S. Department of Agriculture, Washington, D.C., 20250:

Agricultural Marketing Service
- Grain Market News (wk.)
- Feed Market News (wk.)

Crop Reporting Board
- Annual Planting Intentions Survey (end January)
- Crop Production Reports (mo., ann.)
- Stocks of Grain in All Positions (q.)
- Cattle on Feed (mo. and ann.)
- Hogs & Pigs (q.)
- Livestock Slaughter (mo. and ann.)
- Meat Animals (ann.)
- Broiler Report (wk.)
- Eggs, Chickens & Turkeys (mo.)
- Poultry Slaughtered (mo.)

Economic Research Service
Dairy Situation (5/yr.)
Export data (various)
Livestock & Meat Situation (bi-mo.)
Livestock Meat Statistics (ann.)
Demand & Price Situation (q.)

Foreign Agricultural Service
Foreign Agricultural Circular, Grains (irreg.)
Various export publications (irreg.)

Bureau of the Census
U.S. Department of Commerce
Industry Division
Washington, DC 20233
Fats & Oils, Production, Consumption, Stocks (mo.)

Other Sources The various trade associations listed in the Directory of National Trade and Professional Associations which is available in most libraries. Associations are listed by subject of interest. Some examples are:

American Feed Manufacturers Association
1701 N. Fort Myer Drive
Arlington, VA 22209

American Institute of Baking
400 East Ontario Street
Chicago, IL 606ll

American Soybean Association
Box 158
Hudson, IA 50643

Grain Sorghum Producers Association
1212 14th Street
Lubbock, TX 79401

Great Plains Wheat, Inc.
1030 15th Street NW
Washington, DC 20005

National Association of State Departments of Agriculture
1522 K Street NW
Washington, DC 20005

National Association of Wheat Growers
1030 15th Street NW
Washington, DC 20005

National Corn Growers Association
Box 358
Boone, IA 50036

National Grain and Feed Dealers Association
500 Folger Building
725 15th Street NW
Washington, DC 20005

National Grain Trade Council
725 15th Street NW
Washington, DC 20005

National Soft Wheat Millers Association
175 West Jackson Boulevard
Chicago, IL 60604

National Soybean Processors Association
1225 Connecticut Avenue NW
Washington, DC 20036

Transportation Association of America
1101 17th Street NW
Washington, DC 20036

Water Transportation Association
500 Fifth Avenue
New York, NY 10036

Wheat Flour Institute
14 East Jackson Boulevard
Chicago, IL 60604

The Chicago Board of Trade publishes a number of books and other publications on the uses of futures markets and individual futures contracts. To receive a publications list, you may write:

Literature Services
Chicago Board of Trade
LaSalle at Jackson
Chicago, IL 60604

index